Ian Vince is a contributing editor to *The Idler*, has written for Channel 4's *Bremner, Bird and Fortune* and is the author of four other books, including *Britain: What a State* and *The MyWay Code*. Since 2008 he has written a regular column for the *Daily Telegraph* called 'Strange Days', in which he travels around Britain seeking out curious local customs, folklore and odd occurrences. He is also the founding member of the British Landscape Club, which can be found at *www.britishlandscape.org*.

Also by Ian Vince

Britain: What a State

The MyWay Code
(with Dan Kieran)

Three Men in a Float
(with Dan Kieran)

The Little Black Book of Red Tape

Ian Vince

THE LIE OF THE LAND

An Under-the-Field Guide to the British Isles

PAN BOOKS

First published 2010 by Macmillan

First published in paperback 2011 by Pan Books
an imprint of Pan Macmillan, a division of Macmillan Publishers Limited
Pan Macmillan, 20 New Wharf Road, London N1 9RR
Basingstoke and Oxford
Associated companies throughout the world
www.panmacmillan.com

ISBN 978-0-330-53539-7

1 3 5 7 9 8 6 4 2

A CIP catalogue record for this book is available from
the British Library.

Typeset by SetSystems Ltd, Saffron Walden Essex
Printed in the UK by CPI Mackays, Chatham ME5 8TD

To the wonderful Vince girls

Kate, Freya and Eloise

Contents

Contents

Introduction

Britain, it is often implied and sometimes explicitly stated, is the home of the middle-of-the-road and median; the temperate, average and unexceptional. Our climate, though it has its occasional moments of madness, is by and large (and despite the amount of time we spend talking about it), fair to middling. Our plants are quite ordinary; they do not bear luxury fruit and our trees are nowhere near the tallest, broadest or the most verdant on earth. Even our wildlife doesn't seem to be terribly wild – not one animal in these islands could tear your arm off for lunch. Our native mammals, in particular, have all the savage menace of a koala bear in a 14-tog duvet. Of all the sweeping generalisations about our islands, however, the one that seems the most undeniable is that in terms of world ranking our physical geography is unimpressive. Ben Nevis and Snowdon are minor peaks and the River Severn is a mere trickle compared with all the various Niles, Amazons, Danubes, Yukons and Yangtzes. All in all, the received wisdom is that Britain is no more than a foothill to the world.

It was a surprise, then, when a couple of years ago I

suddenly began to feel overwhelmed by the British landscape. I was on a car journey from Cornwall to Wiltshire on the way back from a relaxing family holiday and I had decided to pass the time as passenger by collecting brief impressions of the various landscapes on offer. By the time the car was passing through Dorset, on the hinterland of the Jurassic Coast, I was struck by the dizzying variety of what I saw and a familiar feeling was starting to take hold. Earlier in the year, I had crossed the country with two friends in a 1958 electric milk float at an average speed of around 11 mph (it was an effort to rediscover, like Che Guevara and Fred Dibnah before us, our own country)[1] and – given that this relatively modest journey had taken three whole weeks – had begun to believe that Britain was not only larger but also more interesting than we had ever given it credit for. In fact, I had begun to suspect that Britain was positively exotic. In front of me now was the proof: a bewildering array of landscapes at every turn of the road, each with its own distinctive ecology, microclimate, pattern of human settlement, and all of them dependent on the nature of the materials and structures just under the surface of the land. It was as if I was seeing the country anew. Even though southern Britain was superficially familiar to me, my mental notetaking had turned up millions of years of high drama filled with volcanic mayhem, lush tropical vegetation, polar weather and the most bizarre creatures ever to walk, swim or fly around the planet. Under the tyres of the car, rolling along

1. That is an entirely different story. Please consult *Three Men in a Float*, the book of that particular adventure, for further details.

the A35 between Harcombe Bottom and Puddletown – where limestones, clays, greensand, lias, cementstone and chalk meet up in complex relationships to one another to form an inordinately complicated landscape – was a festival of buxom hummocks, of discordant coastlines, landslips, the odd vertical bedding plane, faults, unconformities and all the rest of it. On top of all of this, and also dictated by it, were the tumuli, stone circles and other ritual landscapes of our forebears who moved on to these islands when the last glaciations of our current ice age receded.

All of this fascinating detail is revealed by the underlying structure of the land and an understanding of how that structure affects human life, but first I should disclose what had suddenly sparked this realisation. I was interested in the subject as a geeky teen, when a general fascination with the natural world and a specific one in birds looked for a while as though it might blossom into a career managing wetlands and wearing Aran sweaters in the teeth of Norfolk gales. I discovered geology as a kind of scientific distraction – an A level to make up the numbers between biology and physical science (the most abstract and complex bits of physics and chemistry combined into one subject for your brain-buggering amusement) – and suddenly the world opened up. Or, rather, the earth opened up and the mechanics of what made the landscape look the way it did became a lot clearer than it had before. I quickly dropped physical science because it was the work of the devil, but geology at least provided field trips – although, to be fair, most of these field trips were to places like railway cuttings and quarries, which only reinforces the impression that the subject has all the sparkly, scintillating glamour of mud.

All of which brings us to this book, which is not meant for the amusement or edification of geologists or landscape archaeologists. If you are a member of either of these esteemed professions, you should put this book down and read something else, because there is nothing here for you. In fact, the contents of this book will, in all likelihood, ever so slightly irritate you. Despite my early interest, I am neither a geologist nor archaeologist – not a scientist or historian at all, in fact; merely someone with an interest in the landscape and who finds scenery an endless source of fascination as well as an inspiration.

The Lie of the Land grew out of that fascination, but also as a reaction against almost all of the literature that seems to exist on the otherwise compelling subject of what made the landscape the way it is. Unlike the fields of botany, ornithology, local history, folklore and countless other domains of the mildly curious, there didn't seem to be any easy way in to the subject of the landscape itself. There are, of course, lavishly illustrated books on gems, rocks and minerals which tend to treat the world as a collection of specimen jars and display cabinets, divorced from the landscape from which those specimens had been collected. There are also a number of beautiful, glossy encyclopedias of the earth, sweeping in their scope, which render knowledge on a continental scale, but which seem incapable of helping you answer even general questions about your local river valley – unless you happen to live on the banks of the Nile, the Amazon or the Mississippi delta.

If you want to know why your nearest river flows the way it does, or why your local hills are gentle on one slope and asthmatically vertiginous on the other, you are out of luck. Worse still, reading the books and journals that try to answer these questions sometimes makes you feel like you'd have to work for an oil company for ten years or write a doctoral thesis on 'destructive tectonic plate margins' to find out. And you didn't really want a career in structural geology, you were just asking a question. You wanted an answer, not a curriculum.

It is not really the fault of academic papers and standard textbooks that they are hard to read and impenetrable for laymen like you and me. They are not written for us, but a specialist audience. There is a paucity of information between generalities about the world and the intricately detailed, peer-reviewed proceedings of academia, between the merely pretty and the rather gritty. What I wanted was a rough introduction – an under-the-field guide – to features in the landscape and how they formed.

It is an all too frequent refrain in the prefaces and back-cover blurbs of books, but in this case, it is true: I wrote this book because I wanted it on my shelves. The aim of *The Lie of the Land*, therefore, is to present the processes that brought the landscape of this country into being, but to do so in a way where the romance of the scenery is not lost under an olistostrome[2] of hardcore scientific language. Where colourful characters – the geologists and landscape engineers who unravelled and manipulated the scenery of Britain from the seventeenth century on – are remembered and where the heyday of romantic appreciation of the landscape of Britain is honoured.

---

2. Sorry, an olistostrome is an underwater landslide on a massive scale.

After a few attempts over the course of writing this book to get plain English explanations for this feature or that, I've generally avoided seeking any advice from geologists or archaeologists. This may seem reckless but if I can work things out for myself, armed only with what turned out to be a largely irrelevant Geology O level from twenty-seven years ago, the first year of a similarly unhelpful A level course, the few interesting bits of schoolboy geography I remember and lots and lots of research, then hopefully it should work for you. I have attempted to filter all of my new-found knowledge through my incumbent ignorance in order to present a simple, yet factually correct, version of the events which made our landscape what it is.

Geology itself is a fascinating study. Its literal meaning of 'study of the Earth' ensures that the science is actually a catholic *mélange* of the most interesting bits of all the other sciences; physics, chemistry and biology all play a part and contribute to our understanding of the earth at its basic structural level. Geology is the wellspring of terrestrial science. Almost everything we know of on the planet evolved from the same origin – the insides of the earth. In 1903, the glory days of scientific discovery, the eminent Victorian geologist Charles Lapworth – who pops up from time to time during this book – compared it to astronomy in its all-encompassing nature:

> Astronomy concerns itself with the whole of the visible universe, of which our earth forms but a relatively insignificant part; while Geology deals with that earth regarded as an individual. Astronomy is the oldest of the sciences, while Geology is one of the newest. But

the two sciences have this in common, that to both are granted a magnificence of outlook, and an immensity of grasp denied to all the rest.

Charles Lapworth
Proceedings of the Geological Society of London (1903)

But it is also a complex science, full of convoluted language that turns even the most superficial study of it into a tricky business. The density of the language, however, is a direct response to the complexity of the matter at hand. Like other scientists and technicians, geologists tend to communicate with one another in a kind of specialist shorthand, which makes the discussion of technical matters efficient by assuming a basic level of expertise. Unfortunately, it is a level of expertise that is (by definition) completely absent from the layman, although we really should know a lot of this stuff – it is, after all, our planet that they are discussing.

Landscape archaeology seems less complicated, perhaps because it is based on the structures that humans have created throughout history and the simple human motivations that are behind them. We intuitivcly know about those motivations because, for all our technological progress, we still understand and experience fear, hunger and safety and a sub-set of us worship when our predilections dictate. We are all, in a sense, already experts in the field, knowledgeable on the forces that drive us to construct shelters, plough land, subsume ourselves in a greater whole and understand our place in the universe. In fact, it is the last of these things which brings us to the book you are holding in your hands now, which I hope will help you not only to unravel our place on the planet, but also understand the forces that unwittingly conspired to bring both the planet and us, a few of its passengers, to this point. Those forces are powerful, often violent and sometimes unsympathetic, unobliging and inconvenient to life and it is a remarkable thing that it has survived for long enough to include us in its matrix. After nearly 4,600 million years, it is only during the last few hundred or so that one species has started wondering why.

IAN VINCE
Salisbury, January 2010

CHAPTER ONE

A Gneiss View

Pre-Cambrian eon – from 1 billion to 3 billion years ago

It was a bright day in early March, the sun was shining and a substantial fall of fresh snow lay pristine under an arching blue sky. I stepped off the train at Lairg, midway along the line that winds its way from Inverness to Wick in the far north of Scotland and wondered whether the postbus would be in the station car park or somewhere between there and Lochinver, lodged up to its headlights in snow.

So I counted myself lucky to be greeted by Donald and his postbus. The morning collection in Lochinver was, by his account, a little hairy – he had to take several runs at the hill out of the village and nearly didn't make it at all – but he got through in the end and by the early afternoon, with a slight thaw under way, we were soon heading off on the final 40 miles of my journey to the top-left-hand corner of Scotland.

I was on a simple expedition, to find Britain's oldest landscape, and I could not have picked a better day for it. For one thing, everything assumes a mantle of unfamiliarity in the snow. The world is turned on its head, with the commonplace all of a sudden rendered ripe and ready for

The Pre-Cambrian Lewisian gneisses

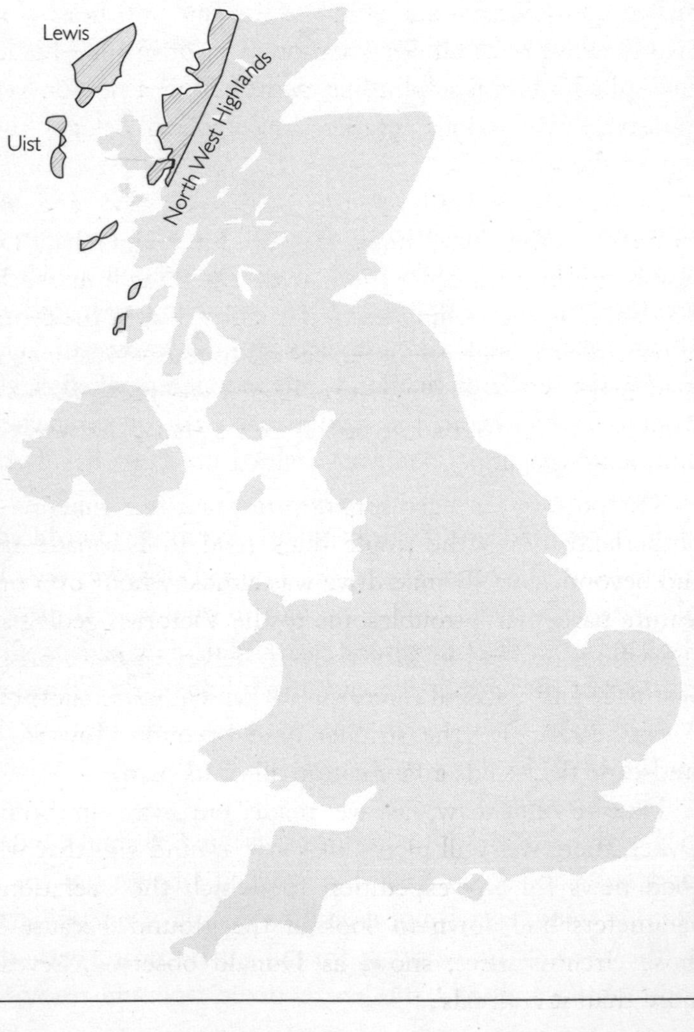

discovery. As a newcomer, with every twist and turn of the road a new experience for me, the layer of snow added dumb secrecy to every moment of the journey, a sense of an approaching surprise, an imminent unveiling, never resolved in its entirety. It also made everything seem quiet, tucked up, quiescent and still; the absolute antithesis of its own creation – of all our creations – a brimstone-fuelled and -filled affair, a sulphurous, acrid hell or a million-year monsoon, the various infernos and biblical deluges that started the world.

Between the wrath of the earth's creation and the stillness of that day, however, the Highlands had not simply ground to a halt quietly over time. Far from it, the Scotland I was travelling to, in particular, was at the centre of enormous turbulence in the earth's formative history and then again over 2.5 billion years later. I was going there on a mission to uncover some of this chaos anew; that was my imminent unveiling.

The postbus barrelled through the ancient mountains of Sutherland, down the single track road to Inchnadamph and beyond. Our 40-mile drive was almost wholly over one feature particularly troublesome to the Victorian geologists and which resulted in fifty years of debate, point-scoring, obstinacy and political careerism (of which, more later) but I was heading for the strip of land beyond – towards a landscape that holds a fascination all of its own.

Despite the thaw, as we made our way up Strath Oykel, there was still plenty of snow around and that was good news for any expedition for which the operational parameters boil down to 'look at the ground' because in those circumstances, snow, as Donald observed, 'reveals more than it conceals'.

He was right in that the overall structure of things becomes more visible when detail is obscured in this way; in such conditions, for example, the thin horizontal terraces – bedding planes – that girdle the mountains are picked out like the lines of ribbing on a gargantuan airship. Reduced to its essentials, snow-bound scenery is an invitation to explore the grand line of nature, the sweeping statement of a landscape rendered by brute force on a planetary scale. If you don't believe me, the next time you experience some severe winter weather, go and look at your local landscape; you will suddenly notice a few details you have never seen before – perhaps the odd hummock or two, or a series of lines on a hillside that reveals how soil creeps slowly downhill over time. It is all there, the detail of it just needs to be hidden in order for us to notice.

As the bus trundled on, a quick glance at the Ordnance Survey map underlined the daunting beauty of the terrain. Feinne-Bheinn Mhor, Coigach, Cul Mor, Quinag and Suilven; the names of draughty peaks came howling from the map like disconsolate wails. Lochs Glendhu, Borralan, Urigill, Culag and Bad a' Ghaill, too, sounded like Tolkein's place names, but the scenery outside the bus was of even more epic proportions. Even a lone stag, magnificently and picturesquely framed on the skyline like a commercial for whisky, water or life assurance, failed to draw my eye for long before it was pulled back to the matter in hand, the potency of the land itself.

West of the road that connects Inchnadamph, Stronchrubie and Knockan Crag, with Durness to the north and Ullapool to the south, that power began to become apparent. Here lies some of the basement rock of Britain – a slab of crust two-thirds the age of the planet on which the entire edifice of Scotland is presumed to rest. This group of rocks – known broadly as the Lewisian gneisses – are metamorphic in origin. That is, they were originally another type of rock altogether and were changed into the form we see today by enough heat and pressure to recrystallise, but not melt, them. They are amongst the oldest rocks in the world and, because they are of metamorphic origin, the Lewisian gneisses were formed somewhere deep, perhaps 30 miles, underground within the crust. From a distance they look a little like granite, but at close quarters they have a distinctive, banded appearance, their alternating dark and light stripes a by-product of their particular formation under extreme heat and pressure. Under the conditions in which the rocks of the Lewisian Complex were formed, minerals that make up the rock were separated out and arranged themselves into alternating layers: the higher density minerals are dark while those of less substance form paler bands within the rock. In places, the stripes twist and turn, squeezed this way and that under monumental force as if the rock once had the structural integrity of toothpaste.

Donald pointed here and there as he told countless stories from the Highlands. Of bank robbers holed-up in isolated shacks; the village bobby turned part-time poacher; and the old lady from Caithness who fought off a military rescue team with a broom during the 1947 blizzard because she believed her rescuers were from another planet.

As Donald stopped to drop off the daily papers at an isolated farmhouse, I saw the chance to take a picture while he struggled with the uncooperative latch of a shed. Climbing out of the postbus, I stood on rock that is close to 3 billion years old, itself just a reiteration of the even older crust that lay here before its mutation into gneiss. When the gneiss was turned into its present form, the earth was much more dynamic; the heat flowing from the planet's core was of a greater intensity than now, largely because there was more radioactivity around – radioactivity which by its very nature, has declined over the life of the earth. The crust that eventually became the gneiss would have been buried by the movement of other bits of crust, gradually building up pressure as it was pushed further down. Under such circumstances, the ground I was now standing on would once have been over 650°C.

The ancient rocks exposed in this narrow strip of land between the main road from Ullapool to Durness and the west coast from Cape Wrath to Loch Broom, as well as on Lewis and other Western Isles, are not all the same however and the term 'Lewisian gneiss' is a bit of a catch-all expression. Even a casual sampling of gneiss across the region will reveal that, rather than a distinct and discrete band of rocks, an entire geological system has been metamorphosed. Muds, limestones, volcanic lavas and rocks formed from molten magma that cooled very slowly underground[1] were the original rocks – or protoliths – and were all metamorphosed in the same manner, but are now counted almost as a single species, despite having differing chemical compositions. And even where you can be sure that you are standing on just one kind of gneiss, it isn't always that simple because the landscape is dotted with further intrusions into it. You can see how complex the geology can be in a location just north of Laxford Bridge, at a spot where the formerly single-track road was

1. Magma is, in essence, lava that has not reached the surface.

improved and widened. When the Highland Council made the A838 wider in 1991, a number of cuttings were made to accommodate the new carriageway, one of which has now become the site of the most fascinating lay-by in the country.

Following the granting of European Geopark[2] status, the whole west coast road from Ullapool to Durness – which includes the A838 – has now been christened the 'Rock Route' by Scottish Natural Heritage and hosts a dozen colourful sets of context boards outlining the geological spectacles en route. The whole Geopark venture is geared towards conservation, sustainable development and education and it is clear from the educational materials that they make available, that Scottish Natural Heritage have something of a popularising streak.

That desire to explain in clear, layman's language is responsible for one of the stop-offs on the Rock Route, the excellent Knockan Crag Visitor Centre near Elphin, where families, students and solo tourists can cheerfully whittle away the best part of the day becoming surprisingly knowledgeable about Scottish geography. It is horses for courses on the rest of the Rock Route, however. A lay-by is not a long-stay attraction and is often only a place to decant tea from a thermos flask or start a short, pointed argument about map-reading, so any educational effort has to cut its cloth accordingly. With a joviality some may find annoying, the Laxford Bridge lay-by has been labelled as the Multi-Coloured Rock Stop. (It must be hard to select

---

2. The European Geoparks Network is a group of thirty-four areas – five of them in Britain – all with highly significant geological heritage on display and all promoting themselves as sustainable tourist attractions.

appropriate language and typography for every passing visitor, so the default setting for promoting public interest in science these days seems to be self-consciously 'wacky' – a style which would not be out of place in a hair-gel advert that features three androgynous teenagers getting out of a lift in fluorescent zoot suits with saxophones.)

Arguments about the presentation of science aside, the Multi-Coloured Rock Stop is an astounding feature. Three different ages of rock can be distinguished on the cutting opposite the lay-by and even the artefacts left behind by the road contractor's rock drills – regular parallel incisions down the rock face that look like a geometric overlay – only add to the effect. Pale-grey gneisses represent the original rock. Swirling across it with the twisted teardrop freedom of a Paisley pattern, a dark brown-black basalt which formed as a dyke[3] as it squeezed through a weakness in the gneiss. Finally, streaks of pink granite cut through both the gneiss and the basalt dyke and so must be the youngest of all three. The gneiss of the North West Highlands feature swarms of later dykes and sills – in particular, there is a group of dykes around Scourie, the exposure at the Laxford lay-by being a particularly good showcase.

Wherever the gneiss outcrops, it does so as open rolling lowland moors, for which the official geological adjective appears to be 'hummocky', punctuated by countless lochs and lochans with the odd mountain towering incongruously above. The hummocks are a product of both the geology and ancient history. Glaciers grinding their way

3. A dyke is an igneous rock formed as magma forces its way through a weakness across the bedding plane of a rock. A similar structure is a sill, which is formed in the same way, but runs along the bedding plane.

over the gneiss sought out every imperfection and weakness in it and by the time that the latest glaciation had reached its peak around 20,000 years ago, the sporadic tectonic activity of almost 3 billion years had led to widespread and wholesale faulting along with swarms of the same intrusive magma dykes that we see on display in the Laxford lay-by today. The ice – up to 800 metres thick at its height during the last glaciation – scoured the landscape making countless hollows of every size, which are now occupied with lochs, lochans and Scottish puddles.

In a sense the old lady from Caithness was right, though perhaps not in the way that she thought. Contrary to appearances, the military rescue team arriving in their insectoid helicopters, wearing jumpsuits and helmets with visors, were from planet Earth, after all, but this part of the Highlands looks as forbidding as the landscape of another world entirely. Almost, at least, but not quite; it is more as though it is an artist's impression of another planet, like one of those quarries where the BBC filmed cheap science fiction series in the 1970s. There are rocks everywhere, some of which are scattered around in a seemingly casual fashion. It is difficult to scatter rocks casually when they range in size from a Calor gas cylinder to a family estate car, so what seems like nonchalant positioning is not so much the work of a bright-yellow JCB than the result of movement by glaciers of staggering proportions a few thousand years ago. This is no quarry for parking the Tardis in; there is something elemental at work and man has had no hand in it at all.

Most of the rock seems to grow from the ground, cropping out from between clumps of poor, peaty earth in rounded pillows that form hundreds of tiny crags and cliff faces, each about ten feet high. It is one of those rare landscapes, in twenty-first century Britain, that is devoid of all human detail and, because of those little crags, carries off the look of another landscape in miniature, like an '00' scale model of the Peak District or a 1:20 recreation of Bodmin Moor.

Once you have arrived in this unique landscape, you may be surprised, as I was, that it feels so ancient, so primeval – as well as alien and elemental. All wilderness has a greater or lesser quality of undomesticated menace about it, as if an absence of human civilisation strips away our higher rational thoughts to let the limbic mind loose, with its twilight programme of instinctive fear unfettered or filtered by logical thinking. A human embryo, it is said, has a passing resemblance to all the evolutionary steps taken since our ancient ancestors flipped out of puddles and ponds onto land millions of years ago. Perhaps pre-human landscapes like these feel so old because we recognise them subconsciously and revert to using our palaeomammalian brains but, either way, the gneiss is a terrain that plays all kinds of tricks to subtly subvert the expectations of the observer. For one thing, it is lowland moor which is in itself a rarity; most of us expect a long climb up into the hills before encountering a rocky, boulder-strewn landscape, but here there is rugged and unforgiving terrain at the same altitude as an East Anglian hill, a gentle eminence: a down of the kind you might not even realise you have climbed. The low altitude of the moor, the wide horizons devoid of familiar scale objects and the peculiar landscape-in-miniature effect also conspire to make any genuinely tall mountain that rises from it into a Himalayan behemoth of the first order.

Near the west coast fishing port of Lochinver, one such mountain the rough shape of a policeman's helmet, Suilven (pronounce *sool-ven*), does exactly that and dominates the entire landscape. It stands like an ancient sentinel and draws your mind to seek comparisons with either a lighthouse or a wild version of Tolkein's 'Eye of Sauron'. Officially classified as an inselberg – an isolated peak that rises abruptly from a plain – the scale of Suilven radiates a magnificent primordial quality over the whole scene. You might half expect, as I did, to see a diplodocus in Loch a' Choireachin, but will have to be content instead with my observations – a pair of goldeneye and a bright-yellow rubber duck thrown in by some local wag.

Suilven, like other mountains that poke up from the Lewisian scenery, is made from almost horizontal beds[4] of Torridonian sandstone[5] that was laid down in large, shallow lakes and rivers when this part of Scotland straddled the tropics a billion years ago. It is just one of a group of similar mountains which run down the west coast between Cape Wrath and the Isle of Skye, a group that probably formed one long escarpment until it was scoured and

4. As an aside, the near-horizontal layers of sandstone show us that not everything is always as it seems. You might deduce from their horizontal structure that they have never been tilted, but a subsequent layer of rocks is tilted up to 20 degrees to the east. Since sediments are laid down in horizontal layers, this means that the Torridonian beds were tilted that much to the west at some point after their initial formation, then eroded to form a flat surface. The subsequent rocks were then laid down and a subsequent tilting has restored the sandstone to its original orientation.

5. Rocks that have been formed under the same conditions and in the same period are named after a defining example called a 'type locality'. In this case, the type locality is near the village of Torridon in Wester Ross, hence they are referred to as Torridonian sandstones.

battered, worn down and washed away by a billion years of wind, rain, rivers and glaciers to its present state.

The inselbergs are therefore the eroded remains of an enormous sheet of sandstone and conglomerates[6] that was once at least 4 miles (6.5 kilometres) deep. Rainfall swept sediment off the gneiss hills and deposited it not only in wide valleys, but also in enormous alluvial fans at the base of the mountains. The erosion and deposition were of such a scale that one alluvial fan is about 30 miles (48 kilometres) wide and almost 500 metres deep (it forms the cliffs of Cape Wrath and extends as far as Quinag[7]). The alluvial fanning pattern of all the Torridonian deposits is most evident in the Bay of Stoer, about a 7 mile (11 kilometres) drive along the narrow and spectacular coast road from Lochinver. Around the northern arm of the bay, the sandstone has eroded along some of its bedding planes to reveal a magnificent sight: the surface of an ancient riverbed complete with a sheet of 1 billion-year-old ripples.

Meanwhile, where all of the sandstone has been removed, it leaves the Lewisian gneiss displaying the contours of a landscape frozen at the moment of its inundation by the sandstone. The gneiss was buried when there were no complex organisms on the planet – when even the diplodocus was a distant and unlikely possibility. In fact, when it was overwhelmed, there was no life at all on land because atmospheric oxygen levels were not high enough

---

6. It's hard to make them sound exciting by telling you that conglomerates are coarse-grained sedimentary rocks made up of rounded fragments (less than 2 mm in diameter) within a finer grained matrix that cements them together, but sometimes the truth just hurts.

7. Pronounced *koon-yag*.

to form an effective ozone[8] layer. Without an ozone layer filtering out the UV radiation from the sun, the surface of the super-continent would have been hostile to all forms of life.

If you look closely around the North West Highlands, you will discover places where it is plain that the billion-year-old Torridonian sandstone lies directly on top of the 3 billion-year-old gneiss. If you do, you will find evidence of an unconformity – a phenomenon which represents a break in the geological record in that, for a period of time, there were either no fresh deposits or those that were laid down were all swept away. In this case, the gneiss has been exposed for a long time before the sandstone was deposited and has been somewhat eroded from its original form as a result.[9] Its exposure to the forces of erosion was long enough to create a landscape which, protected by the shield of Torridonian sandstone, survives to this day. Stripped of what little vegetation grows now on the gneiss, there is no need to imagine the contours of a billion years ago because, between the peaks of the Torridonian inselbergs, they are plain for all to see.

All of this – the inselbergs, the unconformities and the wild, antediluvian overtones – add up to one of the most

---

8. Ozone is an exotic form of oxygen – one with three atoms per molecule rather than the standard two. It forms a high-level layer in the earth's stratosphere at an altitude of approximately 6 miles (9.5 kilometres). The effect of our ozone layer is to shield us from between 93 and 98 per cent of ultraviolet radiation from the Sun. Earth has had an effective ozone layer for only the last 450 million years.

9. At Slioch, a Torridonian mountain that towers above the shore of Loch Maree near Kinlochewe, you can see how a valley in the old gneiss has been filled in by the sandstone.

extraordinary landscapes you are likely to see in these islands. The North West Highlands are a singular landscape, but one which demonstrates as well as any other part of Britain the tapestry of events responsible for the lie of the land.

I needed to write about the North West Highlands first, not just because the top left-hand corner is a good place to start any book, but because this is the area of our islands where the story of Britain started. The landscape of lochan and monadnock (and I'm using the North American name for an inselberg for reasons which will become apparent) around Laxford, Lochinver and Assynt is also an excellent example of one of the key tenets of this book, namely that an unfamiliar landscape need not be foreign; that the exotic can be just around the corner.

However, for the rest of this book I want to turn things around, start from the present day and dig down into our islands to show how the landscape evolved over time. Like all good journeys, however, a detour can be enlightening and it's in this spirit that I offer a chapter on some of the concepts and historical background behind the study of the landscape. You can skip it and pick up the journey at Chapter 3 if you just want to get on with the main story; there's a glossary at the end of the book that clears up any definitions that aren't dealt with en route.

CHAPTER TWO

'No vestige of a beginning, no prospect of an end'

On Geological Time

As anyone who concerns themselves with the fascinating prehistory of our islands would be happy to tell you, whether they are archaeologists, anthropologists, geologists or fossil hunters, it is difficult to get a grip on the expanse of geological time that stretches back to the earliest days of our planet's existence. When our lives of decades amount to such relatively short spans on the earth, even historical time seems unwieldy, so what chance does a period of almost 4,600 million years – the approximate age of the planet – have of making a meaningful impression upon us? Geological time is of such an unimaginable extent that even scientists measure it in a series of multi-million-year subdivisions: in epochs, periods, eras, eons and super-eons, of which the shortest is the epoch.

Eon, era, epoch; all of these terms would be likely to be found in the thesaurus under a master heading of 'absolutely ages' (my dream thesaurus would be fairly informal), so it is a little disappointing to find that an epoch (the shortest) is further subdivided into ages, which themselves can last many millions of years. I looked through the geological literature in vain for even finer partitioning; for instance, as all of these terms are little better than synonyms to non-geologists, how about carving up an age into an even more manageable set of units? Perhaps we could call them yonks? While that thought might strike you as being a little facetious, how can a non-scientist, an ordinary person, make any sense of geological time, when one of its shortest units is an epoch (anywhere between 10,000 and 5 million years)?

Eons, eras, epochs, et al

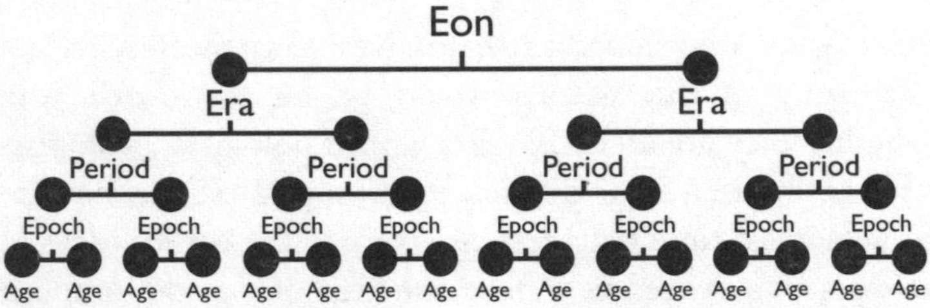

As with any attempt to communicate the scale of something of epic proportions, the numbers soon become so abstract that they lose their meaning altogether and the only way that they can become meaningful again is by recourse to an apt metaphor. In his 1981 book, *Basin and Range*, Pulitzer-winning author John McPhee did just that and explained geological time by using an analogy with the length of the old English yard, a measurement allegedly invented by Henry I as the distance from the tip of his nose to the end of his thumb. On that scale, McPhee wrote, one stroke of a nail file would erase human history. The manicure metaphor is a useful (and literal) yardstick for explaining the puny scale of human affairs when set against the whole history of the earth, but any closer examination of geological time would require a complete understanding of the anatomy of the human arm, shoulder and neck to illuminate the subject fully.

The king's arms notwithstanding, the reductive power of a metaphor can be useful. Large surface areas are almost always expressed as multiples of the area of Wales and large volumes of liquid are framed in terms of the number of Olympic-sized swimming pools that might be filled. With geological time, McPhee's analogy of length is useful, but because we are looking for finer gradations of meaning than are allowed by his single sweep of a nail file that wipes out human history, we should scale it up from a yard to a long journey. As this is a book about Britain's landscape, it seems apt to confine both the timescale and that journey to the landscape of Britain itself.

As we've already seen, our oldest scenery is almost 3,000 million years old. In order to get the best sense of geological time, a very long British journey – a distance of approximately 750 miles (1,200 kilometres) – is the best we can manage without getting on a boat. Which neatly brings us to a scale of 4 million years to the mile and, although it is hard to find a direct road trip of the required length, a slightly scenic route from London to the far north-west of Scotland fits the bill rather well.

As chance would have it, we know that some parts of the area around Cape Wrath, the most northwesterly point of Scotland, were formed around this time, making them the oldest rocks in Britain. At the south-eastern end of the country, there is a lot of geology that is only a few million years old – and even more recent – and the landscape is predominantly capped with very young glacial deposits. London is a convenient choice to represent them: the most recent sediments lie here and there around the M25, down towards the Kent coast and up into East Anglia.

To start with the basics, then: on the scale of a 750-mile journey, a lifetime is equivalent to a journey of a single inch or so, while for the fortunate few who can measure their lives in excess of a century, it is four centimetres from birth to that birthday card from the Queen, the approximate diameter of a toilet roll tube. Everything more than 80 centimetres away – the length of a French stick – is, in a sense, Before Crust (BC/BCE; Before the Common Era, or Christ, if you prefer). The Great Pyramid of Giza was completed approximately 6 feet (1.82 metres) away and, at about 13 feet (3.9 metres) – slightly more than the length of a Ford Fiesta – lies the end of the last glacial period of the most recent ice age, a period of glaciation that lasted for the length of an entire bendy-bus, about 59 feet (18 metres), parked at the far end of the Fiesta. That wasn't the only ice age; there are up to three other major glaciations hypothesised, one of which (on our journey) would extend from Chesterfield to Leeds.

Stepping back a bit further, with regards to Homo sapiens, the consensus is that we probably evolved in Africa just a little further away than the other end of that Olympic-size swimming pool, 52 metres or 130,000 years ago.

For longer journeys back in time, a hypothetical central London starting point at City Hall is helpful. From there it's about 320 metres upriver to get to HMS *Belfast*, a journey that marks the point in time 790,000 years ago that scientists believe our close relatives, Homo erectus, had mastered the use of fire. From the deck of the *Belfast*, we have the extinction of the non-avian dinosaurs[1] in our sights. If you have ever wondered where the guns of the ship are trained, it is – for no adequately explained reason – the M1's Scratchwood Services area, 14 miles (22.5 kilometres) as the shell flies, but 16 miles (25.5 kilometres) or so on our trip via the Edgware Road, representing almost 65 million years. Even the mass extinction of 70 per cent of all species and the end of the age of the dinosaurs – caused or at least aided, it is believed, by the impact of a giant asteroid – hasn't got us as far as the M25 and that is a flavour of just how long the furthest reaches of geological time are.

After our barbecued dinosaur snack at Scratchwood Services, we move further away from London on the M1 to reach the M25 at 100 million years and then Luton at between 140 and 130 million years ago, which marks the evolution of flowering plants or angiosperms – amongst them the oldest fossil of a flower bloom, one apparently related to modern magnolias. By the time we reach Milton Keynes and another famous motorway service station, Newport Pagnell, we are approaching 230 million years ago, a period of time during which it is believed the dinosaurs evolved.

Around Nottingham and Derby, roughly 540 million years ago, is the earliest fossil record of trilobites, the now-extinct marine arthropod. A dozen or so miles north of there, the earliest multicellular animal yet discovered,

1. Birds are now considered to be a specialised group of dinosaurs that survived the mass extinction.

Charnia, pops up from 580 million years ago. Named after Charnwood Forest in Leicestershire, where it was discovered, *Charnia* was found in 1957, without the aid of any kind of magical wardrobe, by a schoolboy called Roger Mason. The journey back to our oldest landscape is only about 20 per cent complete and there are still 600 miles (960 kilometres) or about 2.4 billion years to go.

Heading even further north, the metaphor starts to fray around the edges a little and eventually peters out entirely – mainly because we start to know logarithmically less about the prevailing conditions on the planet the further back we go; there are fewer fossils, fewer rock formations that have remained unaltered and less geological evidence generally.

This lack of detail is reflected in the geological timescale itself. The eon in which we currently live – the Phanerozoic – goes back around 544 million years and is made up of three eras that contain twelve periods which in turn contain thirty-seven epochs. The Proterozoic eon which immediately precedes it is four times as long at 2 billion years, also contains three eras and eleven periods – most of which are based on arbitrary dates, rather than dates calculated by fossil finds or distinctive rock beds (for the simple reason that fossils are few and far between) and lacks epochs altogether.

The line between our present Phanerozoic eon and the Proterozoic is an important one. For a long time, it marked the point beyond which there was no fossil evidence at all. The period that was believed, until relatively recently, to start the fossil record is known as the Cambrian. The sudden abundance of fossilised life at a point around 530 million years ago is called the Cambrian Explosion or

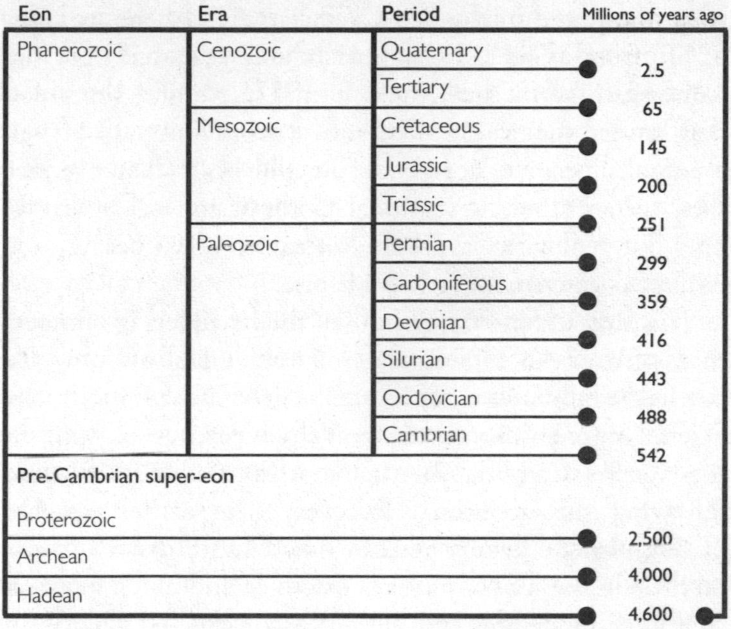
Eon
Era
Period
Millions of years ago
Phanerozoic
Cenozoic
Quaternary
2.5
Tertiary
65
Mesozoic
Cretaceous
145
Jurassic
200
Triassic
251
Paleozoic
Permian
299
Carboniferous
359
Devonian
416
Silurian
443
Ordovician
488
Cambrian
542
Pre-Cambrian super-eon
Proterozoic
2,500
Archean
4,000
Hadean
4,600

Radiation. So profound was this apparent rapid and abrupt diversification of life on earth, Charles Darwin considered it might stand in the way of his theory of natural selection and devoted a whole chapter of *On the Origin of Species* to dealing with it.

To the geologists of the nineteenth century, during which the science flourished in its own radiation of discovery, the abundance of fossils which marked the Cambrian Explosion – compared with the lack of evidence preceding it – drew something of a line under the period. The entire passage of earth's history to that point was formerly classified as Pre-Cambrian, the bottom drawer of the geological cabinet where it was perhaps presumed that answers would never be forthcoming. Eventually, however, more fossils were discovered in the Pre-Cambrian eras (there are signs of life from as far back as 3.45 billion years ago) but the name stuck and is now the informal name of a super-eon that covers the whole 4 billion years. Doubtless, it will eventually become deprecated by the body that oversees such things (the International Commission on Stratigraphy) but geologists still seem content to use it because the term is as convenient as it is historic.

It is easy to forget, amongst all the bewildering numbers that, of over 4.5 billion years of earth's history, only the last half a billion years or so can be accounted for with any degree of scientific certainty. While patches of time in the late Pre-Cambrian become gradually more illuminated following the occasional discovery, the earliest reaches of our planet's history remain much more distant, based on little in the way of physical evidence and much more on computer modelling and elegantly constructed theory. In this book's own bottom drawer, a postscript devoted to the earliest eons of the planet's life, some of these theories are outlined for the sake of completeness. In that postscript, 'A Quick History of the World Before Britain', you will find that the story of what happened during the first 800 million years or so of the earth owe much to a number of serious theories which are based on evidence contained in mineral fragments no more substantial than the width of a human hair.

That scientists can put forward theories based on such physically flimsy trace evidence owes a lot to the principle of uniformitarianism which, even though it sounds like an evangelical belief system, is actually the basis of an entire scientific philosophy: namely that the processes that have shaped the world so far are the same as those that are shaping the planet today. The philosophy of uniformitarianism even reaches out into space with the inference that the universe as a whole is governed by the same laws of physics as those found on earth.

In geology, uniformitarianism is frequently reduced to the helpful dictum that 'the present is the key to the past'. Given the impossibility of observing the processes of the past, the assumption that those processes are the same as they have always been and always will be, started to underscore the scientific thinking of the late eighteenth and early nineteenth centuries.

We have the 'father of modern geology', James Hutton (1726–97), to thank for the concept of uniformitarianism. In his paper *The Theory of the Earth; or an Investigation of the Laws observable in the Composition, Dissolution, and Restoration of Land upon the Globe*, which he and his friend Joseph Black[2] read in two parts to meetings of the Royal Society of Edinburgh in 1785, Hutton was the first to conclude that the earth maintains itself in an infinite cycle.

Of course, if the very gradual processes we can observe now are the same processes that have formed the world, uniformitarianism implies that the earth is very old indeed. Hutton certainly believed so and famously noted that, in terms of the history of the earth, 'we find no vestige of a beginning, no prospect of an end.'

This view was diametrically opposite to the commonly

---

2. Joseph Black was an eminent physicist and chemist himself, discovering latent heat, specific heat, and carbon dioxide. Some of the chemistry buildings at the Universities of Edinburgh and Glasgow are named after him and he was the mentor of James Watt, the inventor and engineer who did so much to improve the steam engine.

held ideas of the time, both scientific and theological. The opposing scientific theory was catastrophism; whereas Hutton believed that gradual changes occurred over long periods, catastrophists believed that sudden, high magnitude events occurred over much shorter spans of time.

The roots of catastrophism were in the dominant ideas of the time, which were nine parts theology and one part geology, perhaps a reflection of early attempts to date the earth from a verbatim reading of the Bible. In his *Annals of the Old Testament, deduced from the first origins of the world* of 1650, James Ussher, the Anglican Archbishop of Armagh, pinned down the start of the world to the night before the 23 October, 4004 BC. The Ussher chronology, as it has become known, arrives at this precise time by totting up all the generations from Genesis onwards.

It is easy to mock the efforts of Ussher in the light of what we know now, but framed in the methods and knowledge of 1650 AD, Ussher's chronology was a serious attempt to solve the mystery of the age of the earth. Amongst all the dates that he worked out for Biblical events was that of Noah's Flood, which he calculated at 2348 BC. The story of a great deluge is, perhaps, one of the most striking fables we hear in childhood, whether it is Noah of the Judaeo-Christian story in the Book of Genesis or the Islamic Noah found in the Koran, whether it is the Aboriginal Dreamtime story of a flood from the mouth of a laughing frog or one of the many sunken land myths like Lyonesse, Cantre'r Gwaelod or Atlantis.

There were a lot of floods in catastrophism – which predicated a large number of high-magnitude disasters over a relatively short span of time – though none of the ages of the earth put forward by the proponents of catastrophism were as short a span as Ussher's chronology which advanced a figure of 6,000 years as the absolute age of the world. Floods were also central to Neptunism, the late eighteenth-century theory that stated that all rocks had crystallised from minerals in the sea. Neptunism's leading exponent was Abraham Gottlob Werner, a deeply charismatic man who worked as an inspector of mines as well as a professor of mining and mineralogy at Freiberg in Saxony. Werner's lectures brought students in from all around Europe, who then returned to their native lands and spread the principles of Wernerian geology as if they were his disciples.

Werner's geology artfully straddled the science and theology of the day – there was even a place in the science of the Neptunians for a good old-fashioned Noachian flood. But by the early nineteenth century, more and more geologists were recruited to the opposite camp of plutonism, which held that the ultimate origin of all rocks on the earth was from volcanic or igneous activity. The concept of plutonism was first proposed by James Hutton and, along with uniformitarianism, it turned out to be the engine of one of his other theories: that of deep time or, as we now know it, the geological timescale.

While Hutton was developing the concepts that underpin much of the modern science, an English mineral surveyor[3] called William Smith was noting the arrangements of rock strata as part of his job for the Mearns Pit coal mine at High Littleton in Somerset and then, a few

---

3. While the early history of geology is populated by curious clergymen and the gentlemen geologists of the landed gentry, mineral surveyors were the first professional field geologists.

years later, as the surveyor for the Somersetshire Coal Canal Company. In the spirit of the times – halfway through the British Agricultural Revolution and at the dawn of the Industrial Revolution – surveying was a boom business. Not only were roads and canals to be built to move coal about the country but the landed gentry and aristocracy were laying out and landscaping the grounds of country estates, whose parks led, in turn, to the notion of an ideal and classically British kind of countryside.

While at Mearns Pit coal mine, Smith had noticed that there was a pattern to the succession in the strata of rocks, all of which could be identified by means of embedded fossils; that each local sequence was part of a universal sequence of strata and that these could be traced for some distance by reference to the fossils contained within. The miners told him that they had individual names for the different beds of coal and Smith soon learned to recognise the differences between these. He was so fascinated by the arrangements of the beds of rock that he was given the nickname 'Strata Smith'.

The mine and the subsequent surveying he undertook for the canal that would take the coal from it to Bath, Bristol and London were the perfect work for Strata Smith. In 1799 he produced the first modern geological map, of an area centred on the city of Bath. He had noticed a map of soil and vegetation type, confined to the same area, in the *Somerset County Agricultural Report*, where its creators, John Billingsley and Thomas Davis, had used an overlay of colours to denote the individual soil and land types. The Billingsley–Davis soil map was a revelation to Smith, and one that he sorely needed. He was already proficient in representing strata vertically, but he was less sure about the techniques

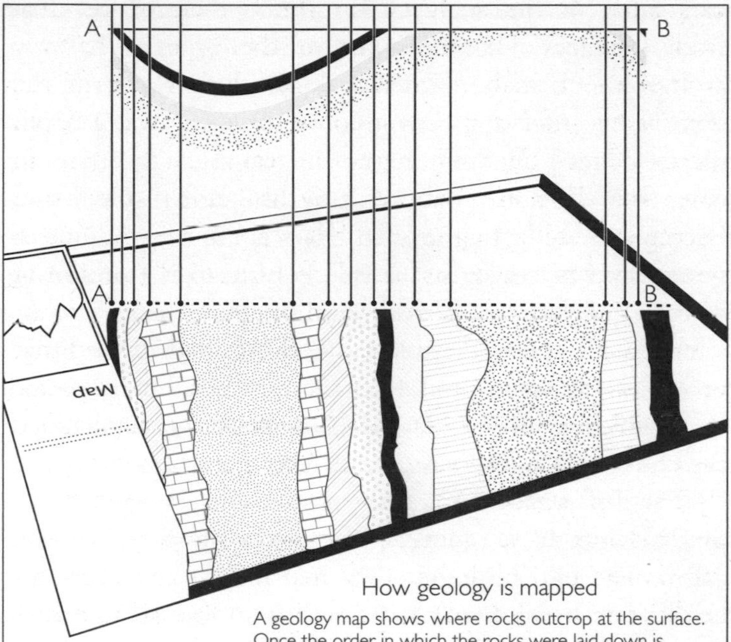

How geology is mapped

A geology map shows where rocks outcrop at the surface. Once the order in which the rocks were laid down is taken into account, the structure beneath the ground – such as through cross-section A-B – can be inferred. (Rear of map obscured for clarity.)

required to represent them on a map. His 1799 Bath map, though unrefined and only showing a selection of rock types, eventually led to his far more polished and now-famous geological map of England and Wales sixteen years later.

While it is true that rocks erode, sediments deposit and continents drift at less than the pace of a snail, the science of geology has raced ahead in the two centuries since Smith, completely unaided, compiled his map. Yet, for all the progress, William Smith's solo effort bears an uncanny similarity to the modern British Geological Survey map, compiled by hundreds of people with the aid of instruments and techniques unheard of and beyond the wildest dreams of surveyors in Smith's day.

If you look at the modern geological map of Britain, it quickly becomes apparent that we live in a geologically diverse country. Resembling nothing so much as the results of an explosion in a paint factory, the British Geological Survey's 25 miles to the inch chart is almost as iconic to generations of geologists as the Tube map is to followers of twentieth-century history and design. Strangers to the science will, however, find it as baffling as William Smith or James Hutton would find the London Underground, but it is only really a question of familiarising yourself with the conventions until you can find your way around.

The different colours of the modern map broadly follow Smith's original selections and indicate the age of what is called the 'solid geology' or 'bedrock'. The bedrock is the consolidated rock that lies under the surface of the earth, beyond the topsoil and the subsoil, as well as any 'drift' or 'surface geology'.[4] The depth at which the bedrock lies varies from place to place. It can be hundreds of feet underground or it can outcrop at the surface.

Geology sometimes seems to be all about rocks and

---

4. Drift is a layer of broken rock debris transported from elsewhere by rivers and glaciers. The distinction is important; bedrock geology of the North Norfolk coast and its hinterland is made up of chalk and some shelly soft sands and clays, but there are glacial deposits that overlie it on a huge scale that you will not see on the bedrock, solid geology maps. These deposits of boulder clay (the name does not need any further explanation) were heaved towards the coastline, by a conveyor belt type process that is inherent in glaciers, towards the snout. In Norfolk, the snout of one such glacier during the last glacial period of the current ice age left behind enough boulder clay to form a line of hills known as the Cromer Holt ridge. At over 300 feet, they are the highest hills in East Anglia.

minerals, their age and their position, but our landscape is much more elaborate than a list of materials from different epochs lying in an assortment of orientations. The chalk of the South Downs, for instance, was laid down between 100 million and 80 million years ago. Over a period of about 20 million years, the microscopic remains of tiny creatures piled up on the bed of a warm sea to a depth of over 300 metres. It wasn't for at least another 60 million years, however, that the seabed with its lithified cargo of chalk along with the clays, pebbles and sands that followed it, was lifted out of the sea by the force of Africa colliding with the southern edge of Europe. That collision left the Alps as its crumple zone, but the effects were felt as far as southern England and north-east France, where the rock layers were gently buckled, like a ruck in a rug, into a 590 feet (180 metre) high long fold. This ruck – which we know as the Weald–Artois anticline[5] – was then eroded down over millions of years to form two inward-facing escarpments – the North and South Downs.

So, the North and South Downs are not just the age of the rocks they are made of. The uplift of the chalk played an important part, as did the erosion that followed. Those two features were just as responsible for the shape of the Downs' escarpments as the marine snow of tiny shells and tests which fell in the warm sea. So it is that the formation of what is rather a simple system is not a one-step process: in fact it is an ongoing process, as is the formation of every acre of our landscape. This is the crux

5. An anticline is an upfold of strata with the oldest rocks at its core. A simple aid to distinguishing between an anticline and its opposite – the syncline – is that an anticline 'points' up like the capital 'A' and a syncline is a 'sink' shape.

Synclines and anticlines

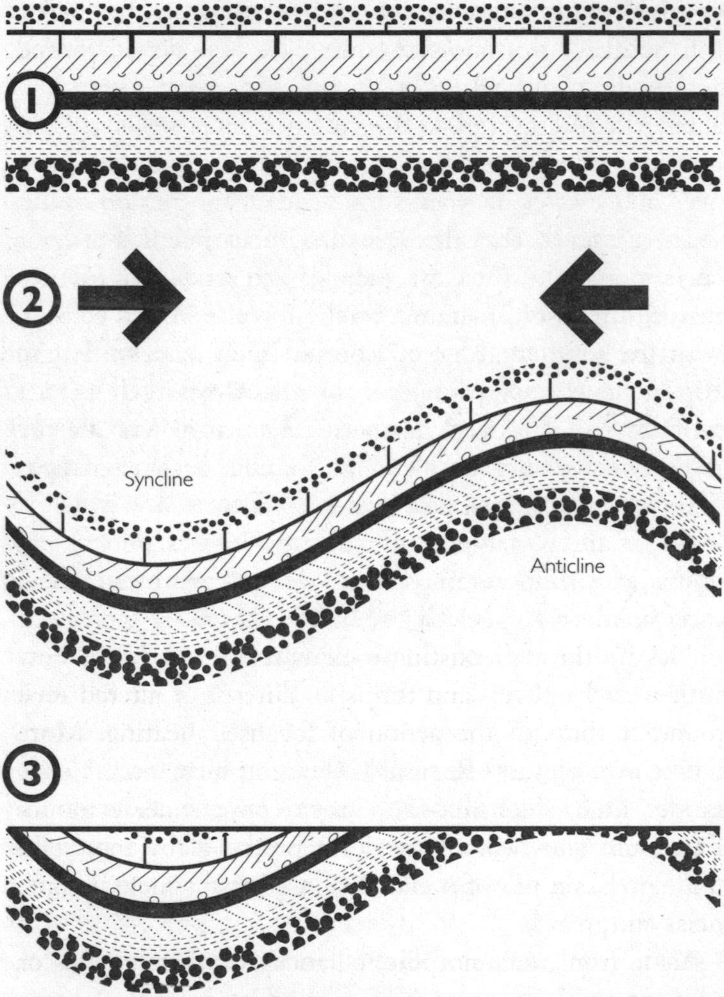

1. Beds are laid down over time.
2. Compression caused by the collision of two plates forces the beds into a form like a ruck in a rug
3. Millions of years of erosion levels the surface

of uniformitarianism: that the present is the key to the past and, as Hutton would say, 'we find no vestige of a beginning, no prospect of an end'.

In terms of this book, of course, we have already found the 'vestige of a beginning' in the shape of the gneiss of the North West Highlands. But even our most ancient patch of ground is, in a sense, older than 3 billion years. Like many ancient rocks, the gneiss is metamorphic, meaning it has been formed by the application of enormous pressure and heat to an even earlier rock. Geologists measure the age of metamorphics from the time they were metamorphosed and were recrystallised,[6] not when the original rocks – or protoliths, to give them their proper scientific name – were formed. The earth was already 1.5 billion years old, time enough for countless reiterations of the minerals and rocks on offer.

There are two types of metamorphism: contact and regional. Contact metamorphism occurs when molten rock meets solid rock, such as when an enormous blob of it bubbles up through existing rock as it did in what is now south-west England, and forms an aureole of altered rock around it through the action of localised heating. More of that in a minute. Regional metamorphism occurs on a grander scale, when masses of rock come under extremes of pressure and heat of the kind produced by mountain building. Examples of metamorphosed rocks include slate, gneiss and marble.

Aside from metamorphic, all rock is either igneous or sedimentary. Igneous rock is formed as it cools from a

---

6. A metamorphic rock is one that has recrystallised under the effects of heat and/or pressure without completely remelting.

molten state like lava or magma. The pumice stone that sits, mostly unused, on the side of your bath is an igneous rock formed when a certain kind of viscous lava is violently ejected from a volcano. Like all lavas, which are rapidly cooled – either by air or by water – pumice solidifies very rapidly and there is no time for its component crystals to grow. As a consequence, volcanic rock is made of fine, tiny crystals.

Not so with the other form of igneous rock, plutonic, named not after a cartoon dog, but the Roman god of the underworld. Plutonic igneous rock cools more slowly deep within the earth and crystals grow to an extent where they can be clearly seen with the eye. Granites, such as those we have just alluded to on the moors of south-west England, were formed like this as swarms of plutons rising from the base of the crust, where they slowly solidified at depths between 3 and 20 miles (5 and 32 kilometres) underground.

Much of the earth's upper crust is made of igneous rocks, but you wouldn't know just to look at it because of the widespread layer of sedimentary rocks that rests on top. Sedimentary rocks (sandstone, shale, limestone, chalk and many others) are formed by deposited sediments, usually – but not exclusively – underwater. Sediments can either be mineral in nature (mud dumped at a river's outlet to the sea, for example), or biological in origin; chalk is formed from billions of tiny shells discarded by single-celled organisms and some limestones are made from coral reefs. The one thing that all sedimentary rocks have in common is that they are formed in horizontal layers – beds, or strata in geological parlance – and build up one on top of another. Because of this simple fact, in any given series of bedrock, the youngest formations are always at the top. It sounds too much like common sense to be the subject of a scientific law, but the boffins like to call it the Law of Superposition, anyway. However, it isn't always that easy to work out where the 'top' and 'bottom' of a geological formation is, because once they have been laid down, sedimentary beds are often tilted – even folded, sheared and inverted, in some cases – by the massive forces that occur when continents drift towards one another.

At the end of this brief tour of the annals of geology, its guiding principles and methods, it is time to turn back to the landscape itself. I started this book in the North West Highlands of Scotland standing in a landscape that was between 1 and 3 billion years old. But even the oldest landscape is, in a sense, still being formed, worked upon by the same forces of erosion that have shaped it so far. Meanwhile, the more recent scenery of southern Britain is merely an expression of the almost eternal process of how the materials of the earth recycle themselves. Having established the parameters, much like the journey we started this chapter with, we are now going on one that takes us from the most recent to the most ancient, digging down through the ages as we go. We take the first steps of this journey along the southern coast of Britain.

CHAPTER THREE

The Unstoppable Tide

The end of the Pleistocene epoch and the Holocene – from the present to 500,000 years ago

You might catch your first sight of it, as I did, from the train as it slows for its final approach to Penzance station. A steep, rugged island, barely 400 yards from the west Cornwall coast and only twenty acres or so in size, rising from the cerulean sea to over 220 feet (67 metres) in height. Connected to the mainland by a rough cobbled causeway at low tide, St Michael's Mount is both picturesque and austere, gritty and pretty – a gargantuan slab of granite and slate that looks good on postcards but oozes with gravitas and authority.

The Mount, as it's locally known, is an established stop-off on the tourist trail of Cornish attractions, its place in the bay assuring it an interesting history and a certain amount of decorative charm – essential prerequisites for any National Trust property. But the heritage colour-way brochures and air of cream-tea loveliness masks an interesting aspect of the Mount – namely, that it is an exemplar of the most recent processes that have formed our landscape.

One good reason for starting a journey here, in the

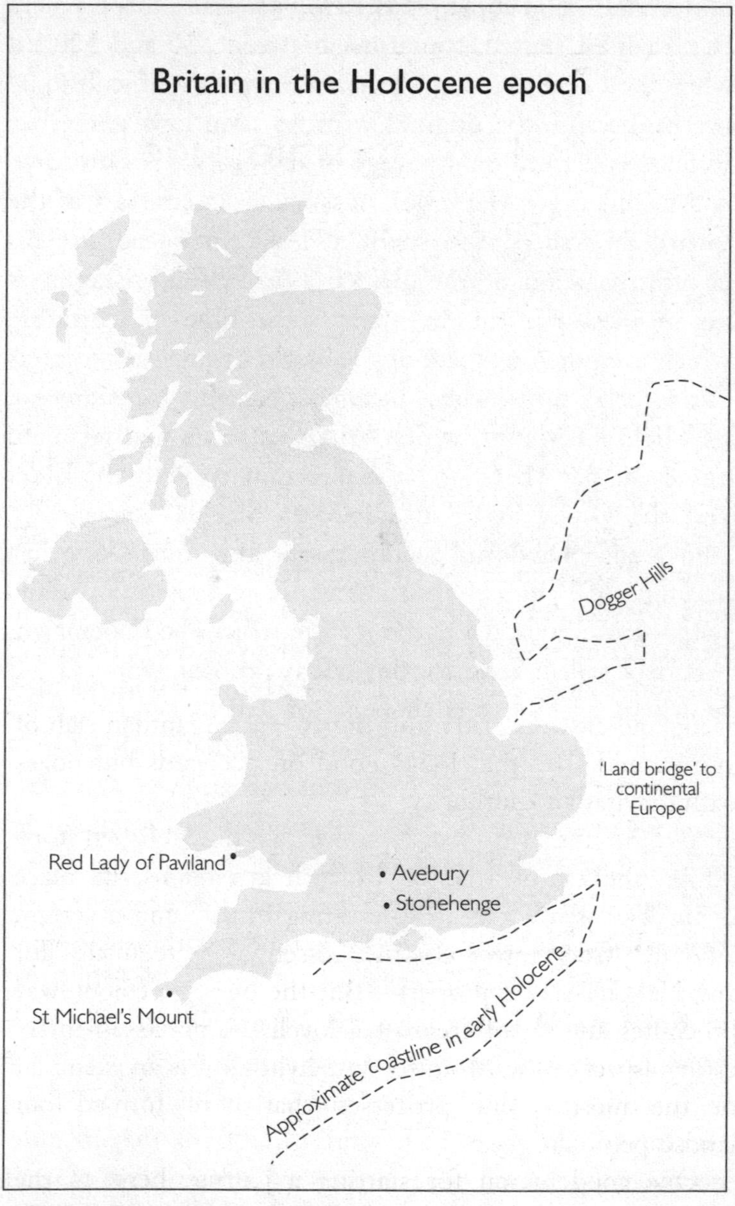
Britain in the Holocene epoch
Dogger Hills
'Land bridge' to continental Europe
Red Lady of Paviland
Avebury
Stonehenge
St Michael's Mount
Approximate coastline in early Holocene

bottom left-hand corner of Britain, is because this is where Britain itself started, sometime between 330 and 320 BC when the Greek explorer Pytheas visited. It's difficult to be precise because the original writings have long gone, but Pytheas embarked on a voyage of discovery from his town of Massilia (now Marseille), possibly as an emissary of the town's tin traders, and spent at least three years on his circumnavigation of the British Isles. Unsurprisingly, as his business was tin, he spent some time in Cornwall, which even then was one of the world's principal exporters of the metal. In particular, he almost certainly spent time on St Michael's Mount, which brings us to its first recorded name – either Mictis or Ictis, according to Pliny the Elder and the Greek historian Diodorus Siculus, respectively. Pliny and Diodorus paraphrased the now long-lost accounts of Pytheas, who first wrote of Ictis as an island with a causeway and a harbour from which the Phoenicians regularly sailed back to the Mediterranean with tin. If St Michael's Mount is the Ictis of 320 BC, which seems likely, it is among the first recorded locations in Britain

But Pytheas's stay in Cornwall had an even more far-reaching and enduring effect – one that has lasted to the present day. At that time, the inhabitants of coastal Cornwall called themselves the Pretani and Pytheas called the British Isles Prettanike. Diodorus rendered this as Pretannia and it was only a short linguistic hop, via the Romans, to Britannia. We may have the Mount and the Cornish coastline to thank for the name of Britain itself.

Impressive as the Mount's claim is to being one of the oldest recorded places in Britain, some think that its most noteworthy achievement is preserving a description of itself from over 4,000 years ago in its Cornish name – *Carrack Looz en Cooz*. A literal translation is 'the grey rock in the woods'. *Carrack Looz en Mor* – the grey rock in the sea – would be far more accurate today, but sea levels rose at the end of the last period of glaciation, at the end of the same Devensian glaciation that was responsible for filling up every ria in south-west England. The result was that a huge area of low-lying forest was lost under the waves.

As if to lend weight to the legend of an extensive forest in Mount's Bay, a number of Cornish tales exist that mention it. On the western side of the bay, there's a patch of sea known as Gwavas Lake which, according to tradition, marks the site of a large freshwater pond situated some miles from the sea and surrounded by trees. Among the more fabulous of the tales is the story of the giant Cormoran, who constructed the Mount as a fortress in the middle of a forest so that he had a clear view over the trees. More conclusively, however, there is archaeological and geological evidence that at least 2,000 years before the earliest form of the Cornish language existed, St Michael's Mount was surrounded by a forest and this forest may have extended as far as 6 miles (9.5 kilometres) to the south into the present-day Mounts Bay.

Today, at very low tides after a heavy storm, the fossilised tree stumps of a forest can still be seen on local beaches. The tree stumps near the Mount have been dated at approximately 4,300 years old. All along the south coast of Britain there are similar fossil forests, briefly erupting from the clutches of periglacial conditions only to be destroyed again a few thousand years later, swamped by the rising tides. The roots of Cornish folklore clung on to the memory of this forest and eventually preserved it in a place name.[1]

Not everybody agrees with this interpretation, of course, but for our purposes it is a useful as well as fabulous story from a time when there was no history, and one that vividly illustrates the impact of the last 2 million years upon Britain. Those 2 million years have wrought many changes to our landscape and most of those changes have been the product of glacial and periglacial conditions. But even in the far south-west, as far away from the ice in Pleistocene Britain as it was possible to go, the glacial periods eventually had the last word when the world's oceans, bloated in the interglacials by meltwater, inundated thousands of square miles around the world, irrespective of climate.

In some parts of Britain, the rise and fall of sea level are complicated by changes in the level of the land. An odd effect caused by the burden of up to 6,500 feet (2 kilometres) of ice is that when the ice was removed, the land 'bounced' back – in fact, Scotland is still rising, while the south-east of Britain, its seesaw counterpart, is steadily lowering itself into the sea.

However you might feel about the Cornish legend of the drowned kingdom of Lyonesse, it is beyond scientific doubt that what is now the English Channel was dry on more than one occasion. The humans who followed the wild game that recolonised Britain after the various retreats of the ice over the last 2 million years would have had no

1. Aside from the name of the Mount, there is also a Cornish Noachian legend of a land called Lyonesse that was lost beneath the waves in a single night as a divine punishment for wickedness.

reason not to colonise the area between Britain and France also. Evidence for a similar tract of land in the North Sea – Doggerland – has also recently been put forward and a few finds have been dredged up by trawlers fishing in the rich waters around the Dogger Bank, thought to be what remains of a range of hills.

During the Pleistocene, the outline of Britain was evolving towards what it is today only with one crucial difference: it was still linked to the continent by a land bridge. This land bridge had intermittently persisted throughout the Quaternary with its frequent glaciations causing long periods of very low sea level. There is now evidence that the English Channel as a topographic feature was first formed as a consequence of two enormous events 450,000 and 200,000 years ago. Using a sophisticated array of sonar data, researchers at Imperial College, London have discovered a deep scar in the seabed which is likely to be the result of a torrent of unimaginable proportions – one of the largest floods ever recorded – ripping from east to west down the channel.

The source of all the water was the southern end of the North Sea, a large glacial lake fed by meltwater along with major European rivers like the Thames and Rhine. At the time this lake was closed to the north by glaciers and dammed to the south by the high chalk ridge of the Weald–Artois anticline (now the North and South Downs) and which originally stretched across the Strait of Dover. At some point, the natural dam was breached and a torrent of 1 million cubic metres of water per second surged down a gently dipping valley towards the Atlantic for a period of several months. It was virtually the same story around 180,000 to 200,000 years ago when a more northerly section of the North Sea became dammed, possibly by a moraine,[2] and was again widening the Strait further.

Without these two events, Britain may have never become an island and the country's, and arguably the world's, history would probably have been very different. It seems curiously apposite that we wouldn't have had the white cliffs of Dover either or be quite as insular (in either sense) as we can be. With the need to defend a land border, our Navy may have been smaller and our maritime ambitions more modest. In short, there might have never been a Britain at all, and we might all live in a region called North Normandy.

Over the length of the Holocene period, the story turns from physical geology to one of human geography. Although the Holocene officially covers the last 12,000 years, there have even been calls by some scientists to recognise the last 250 years as a new period altogether, the Anthropocene, in recognition of the huge impact that humans have had upon the landscape. Just as the incursions of glaciers throughout the Pleistocene have added the final flourish to our scenery and their regression has raised sea levels to form our familiar coastline, the last 12,000 years have introduced these human landscapes to Britain. The 'unstoppable tide' of the chapter title is as much a human tide as a literal one.

Modern humans, Homo sapiens, have been in Britain for at least 30,000 years, although they may have been driven back south from time to time because of an intermittently worsening climate. The oldest modern human

---

2. A moraine is an accumulation of debris deposited at the edge or end of a glacier.

remains found so far in the UK belong to the 'Red Lady of Paviland', actually a twenty-one-year-old male with ochre-stained bones laid to rest in a limestone cave on the Gower Peninsula near Swansea around 29,000 years ago, complete with ivory jewellery and the skull of a woolly mammoth.

After the last glaciers retreated between 10,000 and 15,000 years ago, most of Britain, except a few areas of moorland and other inhospitable habitats, was covered by the ancient 'wildwood'. The wood may have been a little more open on the high and relatively dry ridges of the Downs where the thin soil on the chalk could not accommodate large trees. The woods were cleared for agricultural use by our Neolithic and Bronze Age ancestors, but the soils were too poor to sustain crop growing. The thin soil over chalk did have other benefits for early societies though, driven as we believe they were by a calendar of ritual. Firstly, the chalk is easier to dig than most other rocks, so the excavation of burial chambers, henges and pits could be accomplished with tools fashioned from antlers. Then, once the feature had been dug, it would expose the bright white chalk, which must have been the Bronze Age equivalent of the neon sign. If you think that the hill figures of Wessex (which are mostly the endeavours of eighteenth- and nineteenth-century aristocratic braggarts proclaiming their wealth and power) are impressive, imagine Avebury Henge in Wiltshire, with its enormous bank and ditch, where the difference between the bottom of the ditch and the top of the bank was about 55 feet (17 metres) – the equivalent of a six-storey house, all picked out in white against a verdant down. It must have scared the willies out of passing strangers.

Avebury Henge is interesting because not only is it in Wessex, the cradle of the British Neolithic and Bronze Age, it is also at the focus of an assembly of archaeological wonders – circles, stone avenues and burial chambers, not to mention the largest man-made hill in Europe – that amount to more than the sum of their parts. A proper study of one of these parts inevitably leads to consideration of the others and a dozen or so square miles of Wiltshire and its Neolithic and Bronze Age remains suddenly reveals itself as a single monument, nothing short of a full-blown ritual landscape, where each element becomes a cog in a spiritual machine. There is speculation that the Avebury complex marks the point at which the earlier Mesolithic nomadic lifestyle came to an end. Previously, tribes would range over great distances and honour different deities at different locations through the nomadic, ritual year. When agriculture removed the need to travel, the argument goes, they concentrated their rituals closer to home and constructed a world in microcosm made of parts of the world at large. Many of these specially constructed parts will have been picked out in the white of the chalk, but the henge itself would have been the star attraction – a bank and ditch arrangement that encircled twenty-eight acres and was the result of the excavation of 4 million cubic feet of chalk. It must have been an impressive sight.

Wessex in general and Wiltshire in particular were at the crossroads of many ancient tracks that ran along the tops of escarpments, where the going was easier than the thickly wooded vales. Many of these tracks survive as the various ridgeways of southern England and many of them converge on Wiltshire, which is why the county has an embarrassment of Neolithic and Bronze Age riches.

Around 20 miles (32 kilometres) to the south of Avebury lies the jewel in the crown of this treasure chest: Stonehenge, an ancient monument you may feel you have a passing familiarity with. As an instantly recognisable visual metaphor for the preoccupations and religious manias of our ancestors, the image of Stonehenge is ubiquitous in tourist literature, books on archaeology, postcards, New Age paraphernalia, hippy websites and almost every nook and cranny of British life. It has become a representation of Britain itself, almost as if the familiar red telephone box of Sir Giles Gilbert Scott had somehow become a sacred religious icon, was invested with magical healing powers, was only found at the intersection of ancient and unknowable lines of mystical energy and communication (call-box broadband doesn't count) and a small tabletop version of it was available that facilitated the smoking of marijuana on a personal, yet industrial, scale.

All of this has happened to Stonehenge – although I'm not exactly sure about the availability of a Stonehenge bong. Given its ubiquity, what significance there is at Stonehenge has been gradually ground down over the years and it has become acceptable to regard it as a little bit of a cliché. Which is a terrible pity, because it is truly magnificent on many different levels.

I was lucky enough, one warm summer evening, to be invited on a trip around the stones and the surrounding landscape with Pat Shelley, a local tour guide whose enthusiasm for, and knowledge of, Stonehenge is unsurpassed.[3] I met Pat quite by chance in an empty car park in the

---

3. Pat runs Salisbury and Stonehenge Guided Tours – www.salisburyguidedtours.com.

centre of Salisbury when I was attempting, with two friends, to drive the width of England in an electric milk float. Pat took this barmy idea on board immediately, allowed us to rewire his kitchen temporarily to extract the current and kept us entertained for hours with hot tea and electricity. Impressed by Salisbury, I moved there with my family the year after the milk float adventure and immediately got in contact with Pat.

After a drive north from Salisbury up the Woodford valley, we stopped to look at a tumulus above the tiny village of Lake. Like most chalk landscapes, there are a lot of burial mounds in the Wiltshire countryside, but this one is special because, years after its occupants were safely settled in for a quiet afterlife, the tumulus was enlarged to improve the view of it from the local stately home, Lake House. The Jacobean manor of Lake House – though now the main residence of Gordon Sumner, also known as Sting – was formerly passed down through generations of the Duke family, one member of which, the antiquary the Reverend Edward Duke, published a not so thin volume of work in 1846 entitled *The Druidical Temples of the County of Wilts*.[4]

Although it had long been fashionable to describe all Neolithic and Bronze Age structures in terms of the Druids, Duke wasn't particularly interested in them, except to spend the first six pages of his book explaining the principles of idolatry, all the while carefully drawing out the reader's agreement on its wickedness. Duke then went on to present an astonishing theory that tied together all

4. Available on Google Books, see website www.britishlandscape.org for details.

the major archaeological sites of Salisbury Plain and Wiltshire in a staggering all-or-nothing synthesis. Duke's proposition was that almost the entire landscape of Wiltshire was a kind of sacred planetarium, with seven temples, each one representing a planet. Duke's interminable prose of chapter-length sentences is difficult to edit without the use of explosives, but here are the best of his introductory ravings:

> . . . the planetary temples thus located, seven in number, will if put into motion, be supposed to revolve around Silbury Hill as the centre of this grand astronomical scheme; that thus Saturn, the extreme planet to the south, would in his orbit describe a circle with a diameter of thirty-two miles; that four of these planetary temples were constructed of stone, those of Venus, the Sun, the Moon, and Saturn; and the remaining three of earth, those of Mercury, Mars, and Jupiter, resembling the 'Hill Altars' of Holy Scripture; that the Moon is represented as a satellite of the Sun, and, passing round in an epicycle, is thus supposed to make her monthly revolution, while the Sun himself pursues his annual course in the first and nearest concentric orbit . . . that these planetary temples were all located at due distances from each other; that the relative proportions of those distances correspond with those of the present received system; and that, in three instances, the sites of these temples bear in their names at this day plain and indubitable record of their primitive dedication.

A journal of the day, the *Britannia*, was favourable in its review of *The Druidical Temples of the County of Wilts*, but could not resist sounding a cautionary note:

> He seems both an observant and sagacious man. Yet it is to be noted that the fancy is never more active than when it is engaged in supplying evidences to support a favourite notion. At such times trifling circumstances are magnified into convincing proofs, and the most glaring discrepancies and opposing facts are passed over as beneath consideration. While one set of faculties is called into a state of unnatural excitement, another is lulled into absolute repose.

A local newspaper, the *Wiltshire Trust*, was less kind and wrote a review of the book that started with the line, 'It is a curious fact that lunatics never will believe that they are insane', and went steadily downhill from there.

Duke may have been certifiably, bat-shit crazy and an unlikely early champion of the modern New Age aesthetic of pseudosciences like numerology and sacred geometry, or he may have just been a doddery old country parson trying too hard to make sense out of a widespread scattering of ancient monuments, all of which seemed to have something in common. *The Druidical Temples of the County of Wilts* is an interesting read for all its absurd, spurious claims and magnified trifling circumstances, because it looks at the landscape as a whole, rather than as a series of isolated single sites, and that is by far the best way to approach large ritual landscapes such as those found at Avebury or Stonehenge.

The landscape around all archaeological sites is important. It can tell archaeologists a lot about context and background with regards to the main site, while it is also an area that can provide supplementary material. At large ritual complexes, however, the landscape around the monuments becomes a part of the archaeology of the site; patterns of movement between different monuments may have occurred in strictly delineated ways, using features of the landscape either selected or, in some cases, specially built to manage the emotions of the participants. You can't look at ancient monuments in isolation, they always exist within some kind of landscape context. What would thirty-first century archaeologists make of what will one day be ancient London if they happened to put a dozen trenches across the inner field of the Oval? They might conclude that London was all perfectly flat grassland except for the occasional incidence of tiny post holes, always arranged in a line of three at the end of a ritual walkway. By exploring more of the landscape and plotting relationships within it as archaeologists do now, they would build up a steadily more detailed picture, although explaining the exact nature of cricket would take an anthropologist with more academic chutzpah than the Reverend Duke ever had at his command.

That Stonehenge cannot be viewed in isolation from its landscape would seem to be obvious, but the landscape in question may be larger and more subtly defined than you think. Modern highway building, and its representation on maps, has conspired to make us view landscapes as the interstitial blocks between roads, the white space in the road atlas. The dominant feature in the Stonehenge landscape, as revealed by a glance at any route map of the area, is that of a triangular pennant pointing east towards the town of Amesbury. The pennant is formed by three roads – the A303, A360 and A344 – and on the map, it is only the Neolithic and Bronze Age remains, spattered like grapeshot across the white spaces, that break up the uncompromising geometry of intersecting A-roads, hinting at something rather special on the ground. These three very busy and fast-moving roads restrict our view of Stonehenge because they fence it in both on the map and, for the purposes of exploring the monument as a part of a wider area, on the ground – as surely as the turnstiles, chain-link fences and security guards.

Our appreciation of the landscape is warped by the development of the motor car and the infrastructure required to support it and at ancient monuments it is often worse. Where there is a conservation requirement to restrict access in some way, where catering for a range of visitors – some of whom might not want to stay very long – becomes a necessity, and where profiting from those visitors becomes essential in the name of preservation of the site, we lose much of the original context in favour of the souvenir shop and the car park.

Pat Shelley promised to show me the four oldest things in the Stonehenge complex: the sites of ancient Mesolithic[5] post holes around 30 inches (0.75 metres) in diameter, all substantial timbers and approximately 10,000 years old. Nobody knows what the posts once supported and a variety of suggestions have been put forward from totem poles to part of a very large timber circle. What is known is that the pine posts were there 5,000 years before the first phase

5. The various Stone Ages of Britain are arranged in a geological fashion with an early, a middle and a late phase, respectively the Palaeolithic, the Mesolithic and the Neolithic. They are defined by the development of human technology which itself varied from region to region, so dates are often contentious. In Northern Europe, the start of the Mesolithic officially begins at the end of the last glacial period, which can be anywhere between around 12,000 to 8,500 BC.

of henge construction. Their presence tells us that there was something about this landscape that was special long before Stonehenge. No matter how important in an archaeological sense they are, features like post holes are rarely visually exciting, but I have to admit that being shown the sites of these amazing finds took me far beyond utter disappointment. There in the car park, driven over by thousands of people a week and with precious little information displayed about them, were a few small, white, circular humps painted on the tarmac. They looked, to all intents and purposes, like 1:4 scale models of mini-roundabouts. Given all that I had discovered about the treatment of the landscape of Stonehenge, it seemed inevitable that sites of such profoundly ancient history were buried under a four-pack of cut-price suburban traffic-calming features. What next, I wondered? Maybe a speed hump over a chambered tomb.

In the book that came from my milk float adventure, *Three Men in a Float*, I made some rather rude and unsympathetic noises towards Stonehenge or, more specifically, the rather stinted manner in which it is presented. It lies beside the A303, the trunk route from London to the south-west, and is, by a long chalk, much the worse for it. As a World Heritage Site – every bit as important, historically, as the Pyramids or any of the other wonders of the world – its treatment in the hands of successive governments is a national disgrace. I hope that, so far in this book, I've managed to maintain a tone of dispassionate observation but, as far as Stonehenge is concerned, I say to hell with that. To handle the landscape around one of the most important archaeological sites in the world – a landscape that is as much a part of Stonehenge as the stones themselves – with all the respect accorded to a motorway service area is utterly beyond belief. We are meant to be a nation that, if anything, shows a little too much reverence towards our heritage. None of this is evident from the treatment of Stonehenge.

One wonders what it was like without the traffic. In 1836, John Constable, a frequent visitor over the years to nearby Salisbury, exhibited a watercolour of the monument at the Royal Academy. It was an unusual study completely lacking the formality of alignments we are so used to seeing in the depiction of Stonehenge and he appended the cryptic note: 'The mysterious monument of Stonehenge, standing on a remote and boundless heath, as much unconnected with the events of past ages as it is with the uses of the present, carries you back beyond all historical periods into the obscurity of a totally unknown period.'

Pat and I walked out on Stonehenge Down heading past swarms of burial mounds and away from the crowds towards the Cursus, a large rectilinear earthwork over 1.75 miles (3 kilometres) long and ranging from 300 feet (91 metres) to 500 feet (152 metres) wide, recently dated to between 3630 and 3375 BC. It is a curious fact that, despite being less than a mile from a million visitors a year, only a few hundred make the short walk to see this curious feature, a feature that is every bit as enigmatic as the main attraction.

A cursus is a strange thing: a long, thin platform distinguished from the surrounding landscape by ditches along its entire length and a ditch and bank arrangement that terminates both ends. They yield very little in the way of archaeological finds – some even appear to have been kept consciously 'clean' – and their purpose is, when it comes to the long and short of it, completely unknown.

Looking at the wider landscape, cursuses are usually one feature of a larger ritual landscape. They have often been typified as ceremonial walkways, earthen routes for procession; in short, prototype avenues. An avenue, in an archaeological sense, means a long, parallel-sided strip of land, open at both ends with edges marked either by stone or timber alignments, a low earth bank and ditch or both. Stonehenge has one such avenue accorded the honour of being called the 'Avenue', as if there were no other. The Avenue, bounded purely by bank and ditch, enters Stonehenge at its north-eastern entrance at the Heel Stone and its path represents the key astronomical alignment of the monument – to the sunrise at the Summer Solstice.

Pat took me across the down a short distance from the cursus to the 'elbow' of the Avenue, where it comes in from the east and suddenly turns south-west. From the moment we had left the car park, Stonehenge had always been there, perched at the head of its ridge, but when I turned around to see it from the Avenue, it had completely disappeared behind a low hill, an eminence of chalk that was inconsequential from all other viewpoints except that one. We then walked up the Avenue and had the experience of watching Stonehenge itself rise above the wild flowers, grass and thistles of the down. For me, this set the matter straight: the Avenue is a processional walkway or, at least, something comparable, a route engineered to impress all those who walked upon it – an act of show-business designed to provoke a response.

The Avenue arrives at this elbow after a gentle curve of

a journey of over one 1.5 miles (2.5 kilometres) from the River Avon at West Amesbury and this link to the river may turn out to be an important part of an even larger landscape, an entire Stonehenge landscape which is not confined merely to the henge, Cursus and tumuli of Stonehenge Down. Two miles (3 kilometres) north of Amesbury and about 2 miles east-north-east of Stonehenge, Durrington Walls is situated on slopes that overlook the River Avon, a few miles upstream from the wet terminus of the Avenue at West Amesbury. Durrington Walls is the largest henge in Britain,[6] around a third of a mile (500 metres) across with a ditch 20 feet (6 metres) deep, 50 feet (15 metres) wide at the top narrowing to 20 feet (6 metres) at its floor. Standing close to its southern rim, it is so large that it is indistinguishable in scale from natural landscape, forming a sloping bowl that leads down to a loop in the Avon. An excavation in 1967 within the henge to facilitate the building of a new road[7] led to the discovery of the remains of two wooden circles, the Northern and Southern Circles, of which the Southern was the largest.

---

6. As henges are unknown elsewhere, Durrington Walls could also be called the largest henge in the world. Despite its long association with stone circles, the word 'henge' means an earthen structure that features a circular ditch inside a circular bank. A circular embankment with a ditch running around the outside of it is likely to be a defensive structure. Interestingly, by this definition, Stonehenge is not a henge, as it has a ditch outside its bank, though few would argue it is a defensive feature.

7. I confess to a little bias here and it is difficult to know where to draw the line but, sadly, the grand line of Durrington Walls was vandalised by the imposition of another embankment in 1967, this time to carry the much less interesting A345 across the bowl as part of a road improvement scheme.

In 2005, the Stonehenge Riverside Project, a major five year study of the development of the Stonehenge landscape led by Professor Mike Parker Pearson, discovered a 100 feet (30 metre) wide road leading from the Avon up to the entrance to the Southern Circle, a distance of over 100 metres. Durrington Avenue, as it was soon christened for obvious reasons, is made up of two layers of compacted flint and is the oldest metalled[8] road surface in Europe. Just as the Avenue at Stonehenge is aligned to the Summer Solstice sunrise, Durrington Avenue is set to the sunset. The project also discovered the remains of the only Neolithic village to be found in southern Britain and, despite only exploring a tiny fragment of the henge and its immediate surroundings, believe that hundreds of houses may be in the vicinity.

As a result, archaeologists are starting to entertain the notion that the area of Durrington Walls and nearby Woodhenge may be where the builders of Stonehenge lived and that the landscape around Stonehenge and Durrington Walls may be divided into two sectors: the living and recently departed at Durrington with its wooden monuments of life, and the ancestral spirits residing around the dead rock of Stonehenge. They were linked together by the River Avon, its two connecting avenues and their rites of passage.

The landscape of Stonehenge, then, is much wider than the immediate vicinity of the monument itself, but includes Durrington Walls and Woodhenge, the River Avon and the human act of honouring the dead. This is where the

8. Metalled is a bit of an old-fashioned term for a road formed from crushed rock.

lie of the land becomes more than just a sum of hillocks, streams, woods and monuments, it is where all of those features, both natural and man-made, are imbued with meaning and called upon by ritual to reflect the landscape of the human condition.

Neolithic activity on downland was strongly influenced by the presence of flint. Before metal smelting techniques were adopted during the Bronze Age, flint was used for axes, arrowheads and various tools. It was the hardest material capable of being worked with precision and was of immense importance in Stone Age society. Flint is found only in chalk and a few limestone beds. It has a hard, crystalline structure, is made largely from quartz and is believed to be formed when dissolved sediments of silica (the main ingredient of quartz) find their way into the burrow holes of crustaceans and molluscs rooting around the still soft chalk sediments on the sea floor. Where there's flint, there are worked flint tools. A great game for children is to have them hunt, in that slightly fixated way, for an old piece of worked flint. As an experiment and after almost no effort at all, I managed to find just such a microlith on Cranborne Chase in around fifteen minutes. It may have been up to 8,000 years old in human terms, but 80 million years old or so, geologically.

Beyond the Bronze Age, mastery and manipulation of the landscape evolved into human domination of it but it isn't really the place of this book to discuss features that are not so much archaeological as architectural. Great towns and cities have grown within Britain not only as a result of human ingenuity and industry, but also as a consequence of the raw materials at our disposal, which boils down to the land beneath our feet. At the same time, as human endeavour moved from a primary phase of agriculture, fishing and the quarrying of rocks and minerals that came easily to hand, through a secondary phase of industrial production, we found ourselves ever more abstracted from the earth we stand upon.

In the future, as the Holocene grinds on and sediments continue to be laid down on lake and riverbeds or deposited out at sea, the next major geological event is likely to be the return of glaciers. Even anthropogenic CO_2 emissions will probably be dwarfed by the next glacial phase of the current ice age, which could return within 5,000 years or so. For all our technological prowess, our modern world is wholly contained within a single interglacial period, a mild spell, an Arctic summer, fifty centuries or so of civilisation set against 20,000 centuries of ice in the recent geological past.

CHAPTER FOUR

The Frozen Landscape

Pleistocene epoch – from 10,000 years to 2.5 million years ago

The glaciations of the last 2 million years or so have probably affected the British landscape more than any other geologically short-lived event in the history of these islands. The Pleistocene 'Ice Age' will dog the pages of this book, popping up as an addendum to the geology of all ages, like a last-minute twist in the plot. There will be many allusions to it and how it has changed the rocks laid down in the landscape we see now. Glaciation, like love, changes everything.

If anything, the principles of uniformitarianism – that the present is the key to the past, to use Hutton's words – had become so well established in geology that proposing the idea of a cataclysmic glaciation was equivalent to defending the concept of a Noachian flood. Widespread glaciation was tainted by the discredited idea of catastrophism and consequently took a long time to be fully adopted. By the 1830s a begrudging acceptance of the role that ice had played in the formation of the landscape found expression in the belief that what we now know as the area of glaciation had been under a very deep sea in which

Approximate limit of 'Ice Age' ice

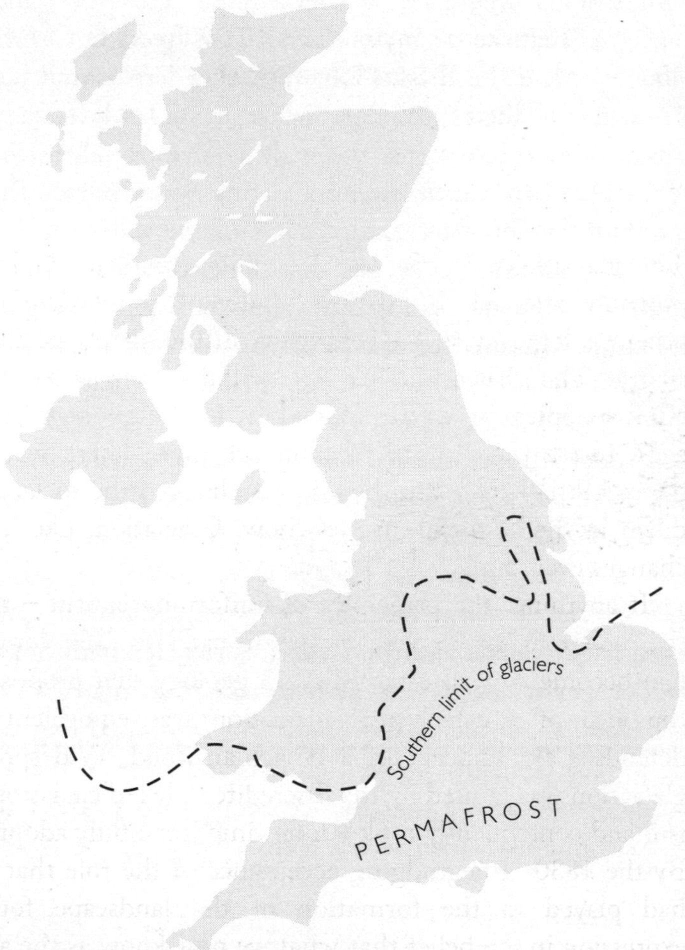

icebergs had floated down from the Arctic. Through studying the effects of glaciers in his native Switzerland, the geologist Louis Agassiz was, in 1837, eventually able to show the world that the extent of ice in prehistory was much more profound than had ever been thought before.

In Britain, where the glaciers rolled across the landscape, and that can be anywhere north of Ipswich, Oxford, Gloucester and the Bristol Channel, they left behind telltale signs of their presence. Most of upland northern Britain shows these signs: the wide U-shaped valleys that are occupied by misfit streams – tiny watercourses that could not possibly have eroded such a large valley on their own; the knife-edge arêtes that formed when glaciers eroded the U-shape on adjacent valleys or where more than one cirque[1] (formed by the heads of different glaciers on the same mountain) meet; hanging valleys where smaller tributary glaciers met main glaciers, often leaving behind a high waterfall once the ice had gone; and heavily striated rock faces halfway up mountains. Most of these features are so easy to visualise as the products of gargantuan ice flows that they form the backbone of GCSE geography field studies in places like the Pennines, Snowdonia, the Scottish Highlands and the Lake District. It's hard not to wonder why science was so slow to take them on.

Beyond the forms that are normally attributed to glaciation – the U-shaped valleys, cirques and arêtes – there is a truly beneficial one that the ice bequeathed to the Lake District: the lakes themselves. While tarns fill the upland

---

1. These ampitheatre-like bowls at the head of a glacier are also known as corries in Scotland or cwms in Wales.

corries, features that are just as distinctive occupy the wide, flat-bottomed U-shaped valleys below: ribbon lakes, each one of them produced as a result of meltwater filling a valley floor scooped out by a glacier. Some of the U-shaped valleys hold not one but two ribbon lakes, separated by flat, green fertile areas that are at a premium for farming in such a landscape and are very recent in origin. They represent the consolidation of deltas formed by tributary streams at the point where they enter the lake; the streams drop their loads of sediments just as a river drops its sediment as it enters the sea to form a coastal delta. Over time the deltas grow and, eventually, one large lake is split into two. The deltas also form at the heads of the ribbon lakes for the same reasons and grow downstream over time. Though divided now and separated by two miles of rich farmland, Derwent Water and Bassenthwaite were once one large lake. Glacial lakes with so much alluvium entering them are essentially temporary features and eventually Bassenthwaite will silt up completely. As will all the lakes over time.

For now, the ribbon lakes *are* the Lakes of the Lake District. Windermere, Ullswater, Bassenthwaite and a dozen more long, dark waters arranged under the fells and pikes in a great radial pattern emanating from Scafell that was first noticed not by a geologist, but by William Wordsworth.

> I know not how to give the reader a distinct image of these more readily, than by requesting him to place himself with me, in imagination, upon some given point; let it be the top of either of the mountains, Great Gable, or Scawfell; or, rather, let us suppose our station

> to be a cloud hanging midway between those two mountains, at not more than half a mile's distance from the summit of each, and not many yards above their highest elevation; we shall then see stretched at our feet a number of valleys, not fewer than eight, diverging from the point on which we are supposed to stand, like spokes from the nave of a wheel.[2]

The drainage pattern itself is a superimposed one that lingers from the period immediately before: the Tertiary, which ran from 65 million to 2 million years ago. During the Early Tertiary, the Lake District was uplifted into a dome with its centre near Scafell Pike. Streams radiated from the top and cut first into chalk, a sandstone and then a limestone, finally incising a path down through shales and slates. The chalk was removed first – as it was over the whole of the north-west – followed by the sandstone and limestone, which now form broken concentric rings around the whole Lake District. Having cut through the older rocks in a manner dictated by the young dome, the drainage pattern has endured, with only a few changes, to the present day.

England's Lake District has inspired poets and artists for centuries from Wordsworth in Grasmere and Coleridge at Keswick to the German émigré and Dadaist Kurt Schwitters,[3] who settled in well at Ambleside from 1945,

2. *A complete guide to the Lakes: comprising minute directions for the tourist, with Mr. Wordsworth's description of the scenery of the country, &c. and three letters on the geology of the Lake District*, John Hudson, Adam Sedgwick, William Wordsworth, 1843.

3. While the exact circumstances surrounding his exile are vague, Schwitters may have said something rather uncomplimentary about the Nazi-endorsed art

probably because of his love of mountain scenery. Between Wordsworth and Schwitters, came another 'mountain man' to live by the Lakes, John Ruskin, the nineteenth-century social critic, artist and philosopher. Ruskin harboured a life-long interest in geology every bit as intense as Wordsworth's – even writing about it, especially in regards to its relationship to architecture. He was not so much interested in the creation of the mountains as the ruins that were left behind; he had little patience for deep time, but was instead stimulated by an artistic understanding of the landscape: 'We have to ask then, first, what material there was here to carve; and then what sort of chisels, and in what workman's hand, were used to produce this large piece of precious chasing or embossed work, which we call Cumberland . . .'[4]

The nineteenth century was an age of synthesis for science and, unfortunately for many, that synthesis did not include a strictly literal or biblical version of Creation. This was problematic for many followers of the new sciences as many amateur scientists of the time had theological training and had come to the subjects almost as a hobby. When inductive reasoning was applied to geological problems, God didn't seem to be present in the detail. Like many, Ruskin[5] may have fought shy of accepting all the scientific presumptions

---

of the time and was wanted for interview – a euphemism if ever there was one – by the Gestapo. He fled Germany in 1936 and his work appeared in the *Entartete Kunst* (Degenerate Art) exhibition mounted by the Nazis to ridicule the Modernists the following year.

4. *Deucalion: Collected Studies of the Lapse of Waves and Life of Stones* (1883), John Ruskin.

5. Like all Victorian polymaths, in drawings and photographs Ruskin's approximate age can be determined by the increasing length of his beard.

of geology, with its impersonal scales of deep time and its increasingly counter-Creation overtones, but he was interested in the landscape and, while a student at Oxford, attended the debates of the Geological Society, including one on the effects of glacier erosion on mountain form. As a consequence, Ruskin's work reads now as an attempt to breathe the mystery back into the landscape, moving dull science aside and reinstating the land with magic.

Eschewing geological time, Ruskin promoted the idea that some geology might be enjoyed for what it is rather than how it formed. 'I do not care, and I want you not to care,' he told his students, 'how crest or aiguille was lifted . . . I do care that you should know . . . in what strength and beauty of form it has actually stood since man was man.' In some cases, however, glaciation has formed landscapes since man was man – particularly during the later glacial periods. Of which, Ruskin would surely approve.

As well as the powerful forces of erosion that glaciers exert, they also deposit the products of that erosion on the landscape. Though its modern name is 'superficial', you are just as likely to hear it referred to by geologists as 'drift' – a hangover from the days when it was believed that icebergs were the chief agent of erosion and deposition. Superficial or drift geology includes a range of some quite substantial forms which I shall give a quick tour of here. All of them are made from the products of glacial erosion: broadly boulder clay or till, a generic name for anything a glacier has ploughed up, partially ground apart and then dumped, unsorted and without any bedding planes. If you live in a part of Britain that was at any time under a glacier, some or all of these forms might be present in your local landscape. Even if you don't live in an upland (and therefore traditionally glacial) area, it's still worth keeping an eye out for glacial features, some of which turn up hundreds of miles from the nearest U-shaped valley.

The most famous of these features are the whale-back hills known as drumlins, rounded hills that often occur in swarms of hundreds with 'blunt' ends that face the origin of the glacier and a long tapered tail on the lee side, very much like the crag and tail structure under Edinburgh Castle and the Royal Mile. The landscape that a swarm of drumlins form is often known by the highly descriptive term 'basket of eggs topography' and can be seen particularly well in Ribbleshead in the Yorkshire Dales or over wide areas of Ireland. Aside from the general principle of streamlining along their long axes on which all agree, the exact formation of these hills is still a subject of research.

Another key feature deposited by all glaciers is moraine, the dredgings of their journey that they push towards their snouts by the conveyor-belt mechanism displayed by all glaciers. The Cromer Ridge along the north coast of Norfolk is one such heap. In an area of the country famed for its low-lying nature, the Cromer Ridge, which is actually the result of two moraines meeting, is a significant feature about 9 miles (14 kilometres) long and rising to a height of 300 feet (92 metres).

East Anglia is covered in drift geology which obscures a lot of the bedrock almost everywhere except at the coast. The region also includes the youngest rocks of our islands, the Quaternary Crags, which, even though they don't contribute any remarkable features to their landscapes, I feel I should mention for the sake of completeness. 'Crag' is an umbrella term for a number of rock types deposited between 3.75 and 1.5 million years ago which range from shelly, pebbly sand to sandy limestones. The oldest of these is the Coralline Crag. Despite its name, which might reasonably lead you to believe that it has something to do with coral, it doesn't – the fossils it contains are of the shared skeleton structures of colonial species of bryozoans. The Coralline Crag is, however, a rare example of a British limestone not produced in a tropical or sub-tropical sea, and is only found in a strip of land less than 2 miles (3 kilometres) wide that runs north-east–south-west for a distance of 11 miles (18 kilometres) from Aldeburgh through Orford Castle, which makes use of the slight eminence of the spot. In this part of the world, even more so than north Norfolk, the 30 feet (9 metres) of relief supplied by the ridge are more than enough to make it stand out, especially when viewed from the River Alde.

Associated with the Coralline but far more widespread is the Red Crags whose chief claim to fame is that they contain the earliest remains in Britain of horses, oxen and elephants. The Red Crags are bursting with more humble fossils of fish bone, sharks' teeth and shells, many of which are from still extant species – the whelks may even look like the ones that can be liberated from a nearby stall, if eating marine snot is your thing.

While glaciers ground down the rock and incorporated it into till, some larger rocks were transported whole, only to be dropped when the ice melted away. These larger blocks – which sometimes appear precariously balanced on the country rock – are known as 'erratics'. In areas of complex glaciation, where glaciers from different areas meet, the erratics can help scientists unravel the sequence of events and what ice came from where. In this way, erratics perform a vital service to science by providing all the utility of a geological homing beacon.

Scientists believe there have been seventeen cycles of glacial-interglacial periods so far and there are more to come. But while a large area of Britain has been repeatedly inundated by glaciers over the last 2 million years, some areas in the south remained glacier free. They were, however, periglacial in nature and subject to permafrost and seasonal snowcaps. The exact effects of tundra conditions on each landscape differed from one to the next.

As anyone who has had to call out a plumber in icy weather will tell you, repeated freezing and thawing of water in a confined space can be very destructive. Water has a very curious property, in that it expands in volume by 9 per cent as it approaches freezing point. That might not seem a lot but, just as it will cost hundreds of pounds to fix the plumbing if your pipes lack sufficient lagging, in nature it is certainly enough to be able to shatter rock. Freeze-thaw weathering is a periglacial effect that can occur in relatively temperate climates because all it takes is the fluctuation of temperatures around 0°C. The high rocks of the Lake District, Pennines, Grampians, North West Highlands and other upland regions of Britain are still being broken into tiny little pieces by frost shattering. These little pieces tumble down the mountains to form scree or talus slopes, the unconsolidated fans of rubble you find at the foot of cliffs and steep slopes in the mountains.

In the south-west, frost shattering has been active on the moors, helping to break up the granite tors by prising apart the rock along its lines of weakness. The process is slow by human standards, but each thaw allows more water into the crack slightly expanded by the ice and so the process goes on.

In porous outcrops, periglacial conditions allow ice to penetrate into the very fabric of the rock itself and change its nature from porous to impermeable. For example, water normally sinks straight through chalk, so surface water and stream erosion are practically non-existent. In periglacial climates, however, it takes on a different set of properties and it is this change in its physical characteristics which leads to a particular kind of erosion.

Although the South Downs were never covered in glacial ice, the climate was still cold enough to maintain a cap of snow and to keep parts of the soil and chalk frozen the whole year round. In the brief summers the snowfields on top of the Downs would melt and the top layers of soil and chalk would thaw and become heavily saturated while lying over the still frozen impermeable chalk beneath. The waterlogged rock slid down the slope en masse over the solid, frozen rock along with the meltwater from the thaw which added river erosion into the fray. As a result of these periglacial conditions, when the ground eventually thawed at the end of the last glacial period, it left behind steep, V-shaped, dry valleys or coombes. Once there was no permafrost left, rain that fell disappeared into the porous chalk once again.

This is how Devil's Dyke to the north of Brighton formed. The most famous dry valley of them all, Devil's Dyke is approximately 300 feet (91 kilometres) deep and over half a mile (0.75 kilometres) long and is a feature that humans have long found so striking, that we have constantly re-invented our association with it ever since it was formed.

Moving out of the Pleistocene epoch, into the present Holocene – out of geological time to prehistory and history – Devil's Dyke was adapted as a hill fort and, in the Iron Age, like many similar sites, it is likely that all of the grass was removed to reveal the stunning white chalk beneath, to either impress or intimidate travellers through, and occupants of, the valley below.

In Victorian and Edwardian times, a mania for tourism took hold. On Whit Monday in 1893, 30,000 people made their way to Devil's Dyke to indulge themselves in the pleasures of what had been set up as a kind of theme park on the site. While a bandstand and fairground would seem to have little to do with bedrock geology, it was the wide open spaces, fantastic views and feelings of freedom that lay so close to bustling Brighton that drew people up there in the first place. It was, of course, inevitable that dull commerce followed them like a Pennine rambler being stalked by an ice-cream van, but Edwardian tourists were lured at first by the products of the geology, even if they didn't know it.

Visitors today need not fear such a garish state of affairs, as Devil's Dyke has been restored to a natural, managed state. But they will be able to see the remains of a 350-metre-long cable car system that ran across the valley, the course of a funicular railway on the northern slope that extended down to the village of Poynings, and the platform bed from a single-track branch line that ran from Hove. It must have resembled the slopes of Snowdon in minature.

The ice has left its mark on the landscape in geological time, prehistory and history, adding to the evolution of Devil's Dyke as it has added to the evolution of Britain as a whole. The last great glaciation, which finished around

10,000 years ago, was a defining chapter in the formation of Britain, topping and tailing the geology and finally rendering the fine detail of the landscape into a shape we would recognise today.

CHAPTER FIVE

The North–South Divide

Tertiary period – from 2.5 million to 65 million years ago

Our landscape seems like a constant and enduring thing; an unchanging, settled backdrop against which life's dramas are played out as if on a stage, in front of fixed scenery. Measured against the longest period we are capable of truly comprehending – a lifetime – the struts and braces that support this scenery certainly seem stable enough. Bits of Suffolk may fall into the sea with alarming haste, we might notice the odd landslip reported on the evening news and we all know that the massed boots of hillwalkers are wearing down the Pennine Way into its constituent grits but, for the most part – as far as Britain is concerned – our landscape seems to change very slowly indeed.

Looking at landscapes on a global scale, the rate of change can seem a lot quicker. Earthquakes ravage the Near and Far East, California, Southern Europe and all around the Pacific Rim, while active volcanoes spit out boiling rock – a feat accompanied, in places like Yellowstone and Iceland, by geysers apparently more accurate in their timekeeping than a Goblin Teasmaid. So it is that, almost every week, we watch tales on television news of cataclysms caused by a fearsomely restless earth, while

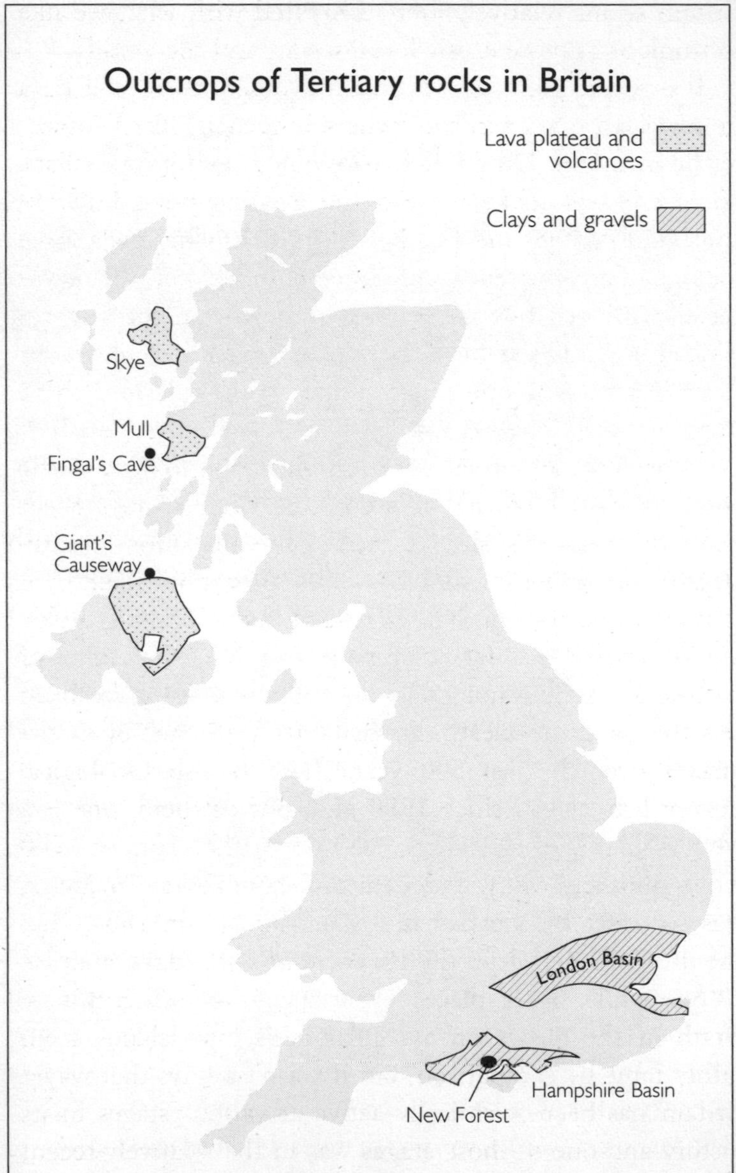
Outcrops of Tertiary rocks in Britain
Lava plateau and volcanoes
Clays and gravels
Skye
Mull
Fingal's Cave
Giant's Causeway
London Basin
Hampshire Basin
New Forest

Britain seems relatively inert compared with what we like to think of as more exotic locations around the world.

It's not all calm, however; occasionally there will be a minor earth tremor in somewhere innocuous like Worcester or Swindon. On an otherwise quiet day for journalists, the event will make it onto the evening news bulletin, consigned to the 'and finally . . .' spot. Voicing over some footage of an apparently unharmed chimney pot, the newscaster will then use the modest numbers on the Richter scale as a gently self-mocking way of saying 'look, even our seismological phenomena are subject to the rules of British irony'.

Indeed, there's not a lot of seismic action to be sarcastic about in Britain, so it's only fair that when it does occur news organisations should extract the maximum smart-aleck value from it, shortly before they film their own outside broadcast unit being crushed under a falling building of course. As far as anyone can tell, that kind of destruction is just not going to happen; there have been less than a dozen deaths in Britain on account of earthquakes over the last 500 years. The British Geological Survey lists eleven since 1580 and, out of them, one – a Mrs Williams of Criccieth – fell down the stairs in 1940 while another, Mary Saunders of Manningtree in Essex, was so upset by the Colchester earthquake in 1884, that she drowned herself in the River Stour a few days later.

So, while other places experience the indiscriminate wrath of the planet on a regular basis, our islands seem rather tame by comparison, but it wasn't always that way – Britain has been seismically active at various stages in its history and one of those stages was in the relatively recent geological past, in the Tertiary period. But before we can establish what happened between 2.5 million and 65 million years ago, it's necessary for a moment to set it against the backdrop of much deeper time.

For example, from 359 million to 430 million years ago during a period geologists refer to as the Caledonian Orogeny,[1] Scotland became joined to the rest of Britain. We'll save the details of colliding continents for Chapter 10, but those 70 million years or so would have had more than their fair share of belching magma and earthquakes at least as ferocious as the worst in recorded history. We are quite used to seeing the grisly scenes of destruction in Japan, the Caribbean and California – all geologically active areas that are the consequence of two or three tectonic plates[2] converging – but the Caledonian Orogeny eventually involved all the world's land masses in the creation of a single continent, Pangaea.

Later, on our whistle-stop timeline, what was forced together in the Caledonian Orogeny was pulled apart again, with the break up of Pangaea during the period[3] immediately before the Tertiary. Gradually the modern

---

1. Although Caledonian Orogeny does sound like at least one potential outcome of watching a Scottish porn film, an orogeny is actually a period of mountain building caused by the collision of two continents.

2. The earth's crust is made up, jigsaw fashion, into a number of pieces, called plates. Large plates underlie major landmasses like North America and Eurasia as well as the Pacific Ocean. Smaller plates underlie the Caribbean Sea and the Arabian Peninsula and all of them move in relation to the others in a kind of global jostle.

3. This is a process that is still ongoing: The Mid Atlantic Ridge is adding distance between Europe and North America at the rate of 25 millimetres a year. Iceland, which straddles the ridge, gave scientists a rare glimpse into the forces that first formed land on our planet when the island of Surtsey emerged as a volcano from beneath the waves in 1963.

continents, moving towards their current configuration, started to emerge.

With those familiar-looking continents moving into place, the world of the Tertiary started to look a lot like the world we know today. Other defining events that drew a line between the Tertiary and its preceding period, the Cretaceous, included the extinction of the dinosaurs, the proliferation of birds and mammals in the niches suddenly left open to them and the creation of the North Atlantic Ocean – albeit in embryonic form. The Atlantic was a direct consequence of the break-up of Pangaea when the forces that were pulling apart Europe, Africa and the Americas created a structure called a rift valley. The rift valley continued to enlarge and ocean water flooded it. Magma flowed up into this rent in the crust, cooled and the result is that the Atlantic Ocean is still being ratcheted apart to this day at a speed of around an inch (25 millimetres) a year.

Meanwhile, in Britain, it would be hard to imagine two more wildly different scenarios occurring at the two geological bookends of our islands. In the south and south-east, two large sedimentary basins were being laid down on the bottom of a shallow sea and in the north and north-west, massive volcanoes belched out lava as the Atlantic began to rift apart.

One consequence of the rifting Atlantic was a network of fractures through which thin, basaltic lava made its way to the surface. The lava, which was runny enough to spread over large tracts of north-western Europe quickly, spread far and wide creating a flood basalt or 'trap' – 700,000 square miles (1.8 million square kilometres) of lava known as the Thulean Plateau. It quickly broke apart and became largely submerged as the Atlantic Ocean widened, leaving large traces of igneous activity from Greenland to the Faeroe Islands and as far south as the island of Lundy in the Bristol Channel.

In the north-west, the Thulean Plateau covered what is now the Ardnamurchan Peninsula, the Inner Hebridean islands of Mull, Rhum, Skye and many others including the island of Staffa. In Northern Ireland, the basalts make arguably their most famous appearance at the Giant's Causeway in the county of Antrim, along the north coast. Its 'discovery' at the end of the seventeenth century and a set of watercolours painted by Susanna Drury, a spinster from Dublin, in 1739, quickly led to claims of it being the eighth wonder of the world. However, readers of the *Radio Times* (who apparently, en masse, know about this kind of thing) decided in 2005 that it was merely the fourth greatest natural wonder of Britain.

The Giant's Causeway is rightly celebrated as one of Britain's most striking landscape features: perhaps because of the polygonal forms of the 38,000 or so basalt pillars which inspire the belief of a human or super-human origin; perhaps because of its location at the sea's edge, which undoubtedly adds to its drama. Or perhaps it is because their form and location combine into a kind of a natural playground for adults and children alike that begs you to clamber over and around the stones to explore them in great detail. There is something irresistible about the pavements of causeway stone, each as neat and geometric as a patio slab, only more interesting, that march out from the base of the cliffs and form short headlands and small bays, one of them, under towering basalt cliffs, going by the name the 'Amphitheatre' for reasons that become obvious to anyone who visits. Because there are over 600,000 people who do just that every year, coming from all over the world, it is easily Northern Ireland's most popular tourist attraction. But the term 'tourist attraction' doesn't seem right in the presence of the kind of gravitas you find at the Giant's Causeway, it just doesn't fit into the theme park mould. People don't so much visit the Giant's Causeway as make a pilgrimage to it.

The regular shapes of the Giant's Causeway 'stones' is a result of the perfect conditions under which it gradually cooled down. To explain it without going into the laws of thermal contraction, think of the shore of a reservoir in a hot, dry summer; as the mud dries out it contracts and cracks in the sun. The columns of basalt were formed in a similar way as they cooled and contracted. The geometric patio effect of the columns only occurs in the middle part of the lava flow; the top and bottom of each flow exhibits innumerable vesicles – tiny bubble holes that were filled with escaped gases like the head on a pint of Guinness. These are less resistant to erosion than the columns and, consequently, are more readily stripped back by weathering to form distinctive, stepped cliffs. The lava fields continue on under the waves across to the island of Staffa, off the western coast of Mull in Scotland, where from aboard one of the little tourist boats that ply their trade around the island, the 'froth' is particularly obvious.

In August 1829, Felix Mendelssohn visited Staffa and was inspired to write the first few lines of what would become the *Hebrides Overture*, also known as *Fingal's Cave.* Mendelssohn once wrote that 'it is in pictures, ruins and natural surroundings that I find the most music' and in the cathedral-like surroundings of Fingal's Cave, with its basalt columns stretching up for 70 feet (21 metres) like great organ pipes towards the cave ceiling, he found the perfect muse.

While nearby Mull may not be graced with its own overture, sonata or symphony, many of the themes of grandeur Mendelssohn worked to such great effect in Fingal's Cave are still present. At Mull, these relatively recent lava fields and all the other related Tertiary igneous activity more or less define the entire island. A fantastic view of layer upon layer of the lava flows can be had from the single-track A849 that runs along the Ross of Mull. The views of Ardmeanach, across Loch Scridain, with its stepped trap landscape are utterly breathtaking. Between eruptions there is plenty of evidence that there were prolonged periods of peace as the basalt was weathered down to make a rich soil. Reddish-brown layers of 'fossil soil' can often be found between the lava flows. Just south of the blunt tip of Ardmeanach – the rounded headland of Rubha na h-Uamha – there is even the cast of a 40 foot (12 metres) high coniferous tree, that was apparently engulfed by lava around 60 million years ago. Named after the man who discovered it in 1811, MacCulloch's Tree can be seen at low tide in the basalt cliff – in the original position in which it grew. Back across Loch Scridain, due south-west at Ardtun, a lake existed at around the same time. Into the lake fell oak, hazel, plane, ginkgo and magnolia leaves, among others, where they sank to the bottom, became trapped under clays and eventually fossilised.

This picture of sedate, if widespread and destructive lava flows with periods of restful peace long enough to grow oak trees and magnolias by a lake, is thrown into sharp relief on the south-eastern corner of the island where the Mull Central Volcano lies – or what's left of it, after almost 60 million years of erosion. It is an interesting choice of name that gives away the underlying truth of the landscape of Mull. We are perhaps more used to seeing something denoted as 'central' when it can be seen as the hub of some kind of human endeavour such as Grand Central Station or the former Central Electricity Generating Board, but Mull Central Volcano is at the hub of one of the most complex areas of volcanic and plutonic activity in Britain.

After the widespread volcanic activity marking the extrusion of countless billions of gallons of basalt lava across the Thulean Plateau, matters became much more centralised as individual shield volcanoes[4] started to develop in the aftermath of the plateau basalts. In Mull, the Central Volcano was one such feature that developed in the area around the present-day Glen More. As the volcano grew after each successive eruption, so did the magma chamber that fed it which, in turn, caused the layers of old lava to dome up under the pressure, folding the rocks into circular ripples around the volcano. The magma exerted so much pressure that it even bends the course of the Great Glen Fault that runs for almost 100 miles (160 kilometres) as a die-straight feature south-west from Inverness before

4. A shield volcano is the shallowest form of what most of us would recognise as a 'traditional volcano' shape. The angle of the slopes is a function of the stickiness of the lava: shield volcanoes with shallow slopes are the product of runny lavas that can run long distances down a slight incline, while progressively stickier and more viscous lavas produce steeper 'composite' volcanoes or 'stratovolcanoes'. Composite volcanoes are a far more common type than the shield volcano seen on Mull.

it passes down the south-eastern coast of Mull close to the volcano. Occasionally, when the magma chamber partially emptied after large eruptions, the central part of the volcano would collapse, creating a huge crater known as a 'caldera'. These calderas could be very large; one that has been plotted from the evidence left on the ground around Glen More is over 6 miles (10 kilometres) in diameter.

The word caldera stems from the Latin *caldaria*, a boiling pot or cauldron, and it is as descriptive of the chemical processes deep within the magma chamber as it is of the form of the crater itself. At first, the lavas were chemically similar to the earlier sheets of plateau lava but, as various minerals crystallised out of the magma, its composition changed over time, a situation further complicated when magma melted the surrounding country rock. This continual change of the raw ingredients of the igneous soup ultimately led to the formation of a pick 'n' mix selection of subsidiary rocks, both volcanic and plutonic. In terms of composition, the rocks that formed from Mull's magma show marked differences over time, but all are just variations around a recipe first brewed by the magma chamber under the Central Volcano.

One of the features that the landscape of Mull and other nearby volcanic centres have inherited from this igneous activity is a repeated circular or arcuate motif. Before a caldera can collapse, a circular fracture called a ring fault develops around the mouth of the cone. These fractures are then intruded with magma from the chamber as the caldera develops and these special forms of dykes are known as ring-dykes. Over the active lifetime of a volcano, this can happen many times and a pattern of roughly

Igneous and volcanic features

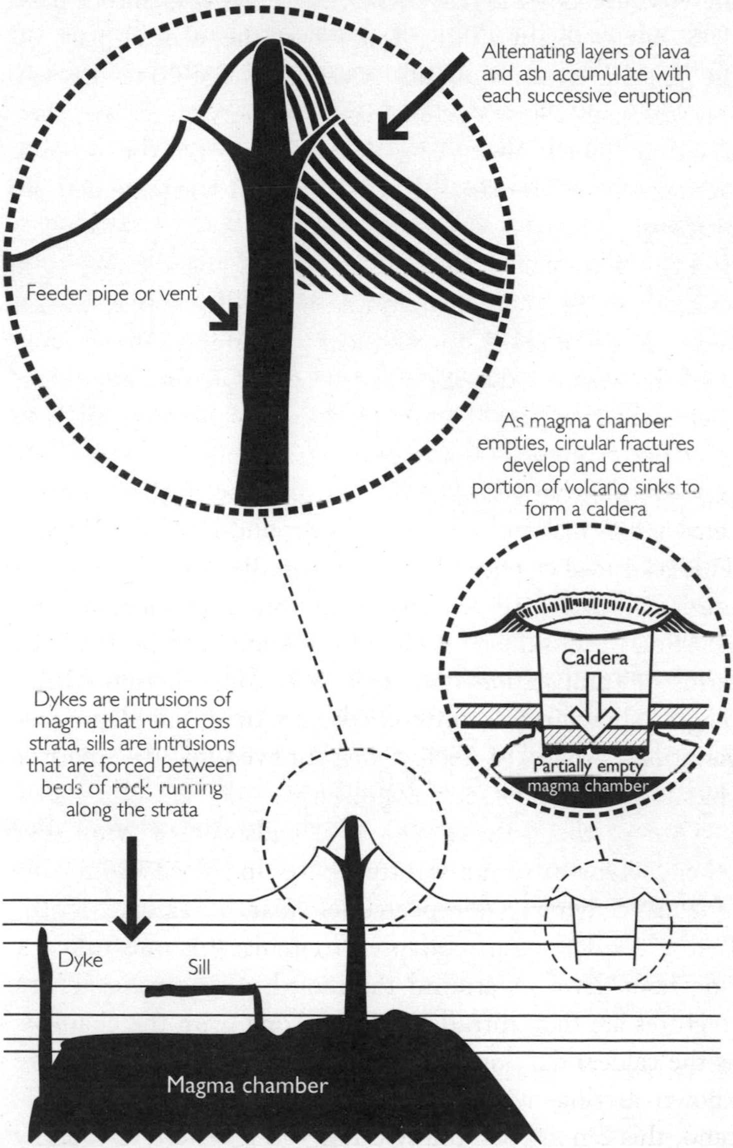
Alternating layers of lava and ash accumulate with each successive eruption
Feeder pipe or vent
As magma chamber empties, circular fractures develop and central portion of volcano sinks to form a caldera
Caldera
Partially empty magma chamber
Dykes are intrusions of magma that run across strata, sills are intrusions that are forced between beds of rock, running along the strata
Dyke
Sill
Magma chamber

concentric ring-dykes that intersect one another will be left behind. Satellite images[5] of the rocky and beautiful wilderness inland of the Point of Ardnamurchan[6] a little to the north of Mull reveal a stunning arcuate pattern formed by a caldera and its associated ring-dykes.

Standing on the Skye Bridge one crisp March afternoon, I looked along the great arching spine of the east coast of the island and reminded myself not to default to the common belief that the oldest scenery is the most rugged. It is an easy mistake to make, especially gazing, at close quarters, up to the snowy, Torridonian peak of Sgurr na Coinnich on the Sleat of Skye. If anything, the reverse can be demonstrated to be true. Older rock is, after all, weathered more and geology of all ages is levelled by glaciers. Skye is perhaps the best place to test those ideas in the whole of Britain because it is where you can find the relatively recent Tertiary igneous formations in close proximity to the rocks of the Precambrian. It is also staggeringly beautiful. The Precambrian rocks such as the Torridonian sandstones, the Moinian schists and – the most ancient of all – Lewisian gneiss dominate the Sleat Peninsula, the face that the Isle of Skye presents to the mainland at Kyle of Lochalsh. All of these have the same time-worn appearance as the rocks around Assynt in the North West Highlands. But a short journey from the bridge up the A87 over Torridonian sandstone along the east coast will

5. Obtained via Google Earth. Placemark files are on the website www.britishlandscape.org.

6. The Point of Ardnamurchan is not a philosophical query on the usefulness of a Scottish peninsula but is, contrary to popular belief, the most westerly point of mainland Britain – not Land's End.

bring you forward over 750 million years as you pass Skye's airport. Another 5 miles (8 kilometres) and you are in the Tertiary Red Hills of pink granite and, between them and the wild west coast, is the Black Cuillin, a short range of mountains also of Tertiary age made from the plutonic and therefore coarse-grained version of basalt, gabbro. Its rough texture makes gabbro a climber's favourite and, given that all of Skye's dozen Munros[7] are in the Black Cuillin, it is a mountaineer's paradise.

Skye's volcanic history is similar to Mull and here the ring-dyke can make matters both easy and tricky for the strangely fixated Munro baggers.[8] On the one hand they can provide advantageous ledges, on the other they provide the most inconvenient Munro of them all, one that has the splendid name of the 'Inaccessible Pinnacle' or 'In Pinn' if you are too busy collecting mountains to speak properly. The Inaccessible Pinnacle is a 150 feet (50 metre) prominence of a ring-dyke inconveniently pointing at the sky and forming the summit of the Munro known as Sgurr Dearg. It is the only Munro that has to be climbed using rock-climbing equipment. Made of basalt, it also tends to be slippery when wet. In his 1947 book *Mountaineering in Scotland*, W.H. Murray commented that the Inaccessible Pinnacle is a 'knife-edged ridge, with an overhanging and

7. A Munro – named after Sir Hugh Munro who was the first to bother publishing a list of them – is a Scottish mountain over 3,000 feet (914 metres) in height.

8. A Munro bagger is a climber who 'collects' the peaks of mountains to no further end than personal satisfaction and in order to develop a wilful disregard of what constitutes an interesting topic of conversation. They can be taken as living proof that some people should 'get out less'.

infinite drop on one side, and a drop on the other side even steeper and longer'.

While the volcanoes were belching out magma in the north-west of Britain, the west of the country was being lifted up as part of the development of the North Atlantic Ocean. Raising the height of streams and rivers adds energy to them and the processes of erosion, which led to the removal of much rock over Wales, the Lake District and Scotland, the sediment flowing eastwards towards a rapidly deepening North Sea Basin. Drainage started to assume an easterly or southeasterly direction and many of the large rivers that drain into the North Sea settled into their current pattern, although many changed their courses following the last glacial period.

Much of what is now Britain was dry land in the Tertiary, but southern England was periodically flooded by what is now the North Sea. Two areas in particular collected sediments which still affect the landscape today; around the London area, down the Thames estuary and up to the Suffolk–Essex border from Sudbury to the coast at Harwich and Felixstowe; and along the south coast, roughly from Bognor Regis to Poole Harbour and inland to the New Forest and east Dorset.

The New Forest, as tired geography teachers must tell their students year in, year out, is not in the least bit new. Or a forest. At least, not in the modern sense of the word. The original meaning of 'forest' really denotes uncultivated land, a definition through which the sense of a wooded hunting area derives. It was the Normans who brought the concept of Forests[9] as the king's hunting grounds with

---

9. I've capitalised Forests because it distinguishes it from the more generic and

them when they invaded in 1066 and the *Nova Foresta* was the only Royal Forest to be mentioned in any detail in the Domesday Book of 1086, by which we can surmise that it is actually among the oldest of Forests.

In order to study the landscape of the Forest it is therefore best to set aside the modern meaning of a forest as an area of woodland, because the New Forest is far from it. In fact, its landscapes can be divided up into three main types, only one of which includes woodland areas. The most prominent of these landscapes are the gravel terraces and sandy plains of the highest ground of the Forest. These plateaux are given over to heathlands on poor soil, with their attendant vegetation of gorse bushes, Scots pine and birch trees. On my subjective assessment, of all the Royal Forests in this country, it is probably the wildest and most interesting. Despite its trifling spot heights, which never exceed 550 feet (167 metres), the heathland of the New Forest has that special kind of desolation that only a moorland plateau can provide; a wide angle view of the world rarely felt beneath 800 feet and aided by the thin infertile soils that have probably better protected it from the ravages of intensive agriculture than any age-old royal warrant.

It is a particular attribute of the New Forest plateaux that the gravels and sands lie directly over clays, at an average depth of 3 feet (1 metre), and that the make-up of the soil also results in impermeable layers which impede the porosity of the sand. This means there are many small pools and standing water can be found on the plateau tops at many times of the year.

---

modern meaning of the word – the difference between a collection of trees called a forest and the New Forest, Forest Lawn, etc.

Similarly, the lower-lying areas of the Forest also suffer from drainage problems and there are large areas of marshland and sedge around the lower reaches of rivers, with carrs[10] of alders occasionally so impenetrable that they remind the casual, water-logged observer[11] of nothing so much as a kind of temperate mangrove swamp, which in ecological terms is exactly what it is.

Finally there are the scattered woodlands of oak, beech, yew and holly, trees which grow on well-drained clays and loams, often on the gentle slopes of the valleys of brooks and streams where erosion has worn away the more problematic strata. Originally, lime trees abounded in the New Forest as well, as evidenced by the name of the 'capital' of the Forest, Lyndhurst, which means 'wooded hill with lime trees'.

North of the New Forest, on the Marlborough Downs, another Tertiary deposit outcrops in an unusual way and is often overlooked, despite contributing greatly to the landscape of Wessex. Some of the sands and gravels of the Tertiary were hardened by silica-rich groundwater to form a kind of hard quartzite sandstone known as sarsen. Over time, the unsilicified sand was eroded away leaving huge chunks of sarsen lying on top of the chalk of places like Fyfield Down where, it is believed, the sarsens of Avebury and Stonehenge may have been transported from.

During the Tertiary, part of the south coast of England would have looked very different from today because the Solent, which is now the inland seaway between the mainland and the Isle of Wight, would have been a large

10. A carr is a wooded swamp.

11. You may have to walk through a stream to get through an alder carr.

river – the result of the confluence of the Frome, the Dorset Stour, the Hampshire Avon and the Test, Itchen and Hamble. The Solent River as it is called, would have been a low-lying, braided watercourse; an estuary at times of high sea level and a non-tidal river with a flood plain when sea levels were low. As we have seen, all of that changed at some point during the Quaternary period that followed the Tertiary, the outcome of a rise in sea level that affected the whole of the south coast of England. But something else happened in that flood. The confluence of half a dozen rivers, the Solent flowed by virtue of a long ridge of chalk that extended from Ballard Down on the Isle of Purbeck, north of Swanage, to what is now the Needles on the Isle of Wight. The story of how this high ridge of chalk came to be there in the first placc, before it was eventually destroyed and washed away, is – appropriately enough – one that begins out at sea.

CHAPTER SIX

A Matter of Life and Death on the Downs

Cretaceous period – from 65 million to 146 million years ago

In 1857, Lieutenant Commander Joseph Dayman was dispatched by the Admiralty at the helm of HMS *Cyclops* to take deep-sea soundings in preparation for the first transatlantic telegraph cable. The laying of this cable, which ran between Ireland and Newfoundland, was as monumental an undertaking as any of its time and, inevitably, its completion was accompanied by lavish celebrations and rambling tributes to engineering excellence. A special committee organised a day of events in New York that featured church services, speeches, a one-hundred-gun salute, firework displays, a 4-mile-(6.5 kilometre-) long procession of 15,000 New Yorkers and the illumination of public buildings. City Hall, indeed, became illuminated in quite the wrong way when a stray firework set fire to the roof, but even that didn't seem to dampen the enthusiasm for celebrating human ingenuity and endeavour.

Unfortunately, the party turned out to be a little premature and the magnificent achievement an all-too brief triumph of science over adversity. Flickering into life on

Chalk downland in southern Britain

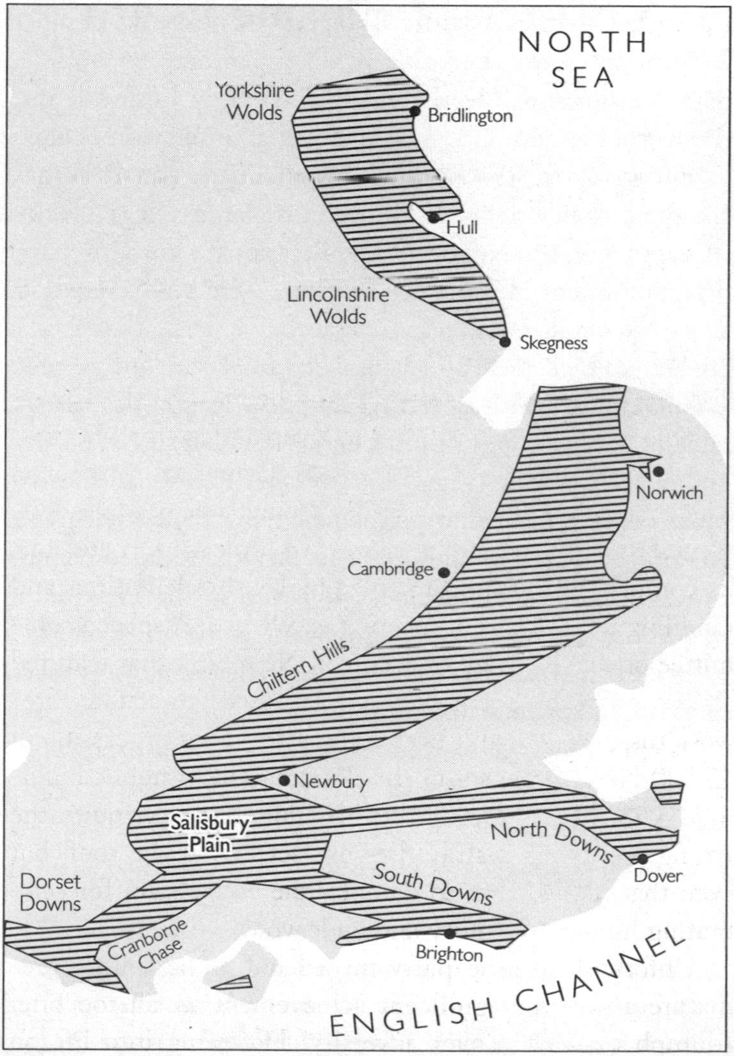

the 5 August 1858, the connection lasted long enough for Queen Victoria and James Buchanan, the American president, to exchange pleasantries in Morse code but was burnt out barely a month later when Wildman Whitehouse, chief electrician to the Atlantic Telegraph Company, pumped 2,000 volts down the cable in an experiment to improve data transmission speeds and inadvertently turned it into the world's largest and most ineffectual immersion heater.

So great was the public excitement generated around the first, albeit briefly, working transatlantic telegraph that its rapid failure caused the exhilaration to turn to bitter disappointment in short order. There were even allegations that the whole affair was a hoax or, at best, that it was cooked up by cynical stock market speculators out to make a quick buck. With potential investors as sceptical as the general public, it wasn't until eight years later in 1866 that another cable, this time laid by the SS *Great Eastern*, became the world's first permanent transatlantic telegraph. Memories of the first failure, it seems, had receded enough to ensure a great hullabaloo. However, in the intervening period, in the much less flamboyant world of geology, a great deal had already been made of the soundings taken by Captain Dayman aboard the *Cyclops*.

Dayman had forwarded specimens of mud brought up by the *Cyclops* to Thomas Henry Huxley, the eminent Victorian scientist who wrote of the mission in years to come with characteristic humour:

> The Admiralty consequently ordered Captain Dayman, an old friend and shipmate of mine, to ascertain the depth over the whole line of the cable, and to bring back specimens of the bottom. In former days, such a

> command as this might have sounded very much like one of the impossible things which the young Prince in the Fairy Tales is ordered to do before he can obtain the hand of the Princess. My friend performed the task assigned to him, without, so far as I know, having met with any reward of that kind.

Huxley didn't know it at the time, but he was to examine mud that would not only start a scientific discourse on the nature of the ocean floor, but also an explanation of the formation of chalk – an explanation that, 150 years later, hints at how the environment may dynamically manage itself in a closed loop, a grand cycle of cause and effect involving a host of organisms that jointly contribute to maintaining a kind of ecological status quo. In today's scientific circles, these mechanisms are at the centre of our current apocalypse fixation – that of climate change.

Huxley examined the samples of mud – which he referred to as 'ooze' – obtained by Dayman and found that chemically it was 'composed almost wholly of carbonate of lime'[1] and that a section of the ooze placed under a microscope revealed 'innumerable globigerinae imbedded in a granular matrix'. Globigerinae is a kind of plankton, a single-celled animal that lives at or near to the surface of the sea and grows a small shell – or test – around itself.

Examining the 'granular matrix' in which the globigerinae were embedded, Huxley noted that all its constituent particles seemed to have a definite form and size.

---

1. Carbonate of lime is what scientists of the day called what is now known as calcium carbonate – $CaCO_3$.

> I find in almost all these deposits a multitude of very curious rounded bodies, to all appearance consisting of several concentric layers, surrounding a minute clear centre, and looking, at first sight, somewhat like single cells of the plant Protococcus; as these bodies, however, are rapidly and completely dissolved by dilute acids, they cannot be organic, and I will, for convenience sake, simply call them coccoliths.

Twenty years before HMS *Cyclops* had even set sail, the German scientist Christian Ehrenberg had noted 'crystalloids' of chalk in samples collected from the white cliffs of Dover. He described them as flat discs, a description that was improved upon in 1861 by the great Victorian polymath Henry Clifton Sorby who published a paper, *On the Organic Origin of the So-Called 'Crystalloids' of the Chalk*. Sorby had ground down a sliver of the rock to a thousandth of an inch and looked at it under a microscope – a revolutionary technique at the time which Sorby had pioneered – and found that 'they are not flat discs, as described and figured by Ehrenberg, but . . . concave on one side, and convex on the other, and indeed of precisely such a form as would result from cutting out oval watch-glasses from a moderately thick, hollow glass sphere.'

While Sorby was peering into his microscope, army surgeon Dr George Wallich was seconded by the Navy on the recommendation of Huxley to take soundings and samples for another telegraph cable under the Atlantic. Although the cable was never laid, the survey nevertheless proved very useful to science. Wallich sent a copy of his report to Huxley which confirmed the existence of the coccoliths that Huxley had discovered but also noticed spheroidal accumulations of them, which he termed 'coccospheres'. Wallich linked the coccoliths and coccospheres to those found by Sorby. Between them, these four men had proven that chalk is almost entirely made out of fossils, albeit very small ones – or calcareous nanofossils as they are known today. To use Ehrenberg's words, chalk is 'nothing more but a heap of skeletons'.

While Sorby, Ehrenberg and Wallich are counted among the leading lights of science of their times, Huxley was, without doubt, a brilliant, talented man, whose work ranged, in that uniquely Victorian way, over a wide variety of disciplines not confined to science. He was the first man ever to coin the term 'agnostic' to describe his lack of religious beliefs, he sat on numerous Royal Commissions on subjects ranging from deep sea fisheries to vivisection, was a journalist and passionate supporter of adult education and fitted all of that in while being the foremost comparative anatomist of his day.

All of which is extraordinary when you consider that he was almost completely self-taught. Mark Twain, a contemporary, famously once said that 'I never let my schooling stand in the way of my education', but Huxley barely received any schooling at all. After only two years at his father's school, his education came solely from reading the latest works of natural philosophy that were in vogue at the time – an autodidacticism that he shared with many Victorian scientists and engineers including Michael Faraday and Alfred Russel Wallace, among others. The son of a middle-class teacher who had fallen on hard times, Huxley left the school at which his father taught at the age of ten, though he didn't seem to regret it in later life, recalling in an autobiographical note in his *Collected Essays* that the teachers 'cared about as much for our intellectual and moral welfare as if they were baby-farmers'.

None of which stopped him for a moment: he published his first scientific paper at the age of twenty – a layer of cells in your hair follicles is named after Huxley – and promptly joined the Navy as assistant surgeon aboard HMS *Rattlesnake*, where he met Dayman. By 1868 he was an eminent and respected scientist and had even earned the sobriquet of 'Darwin's bulldog', because of a notorious moment in defence of his friend Darwin's theories of evolution.

Details are sketchy and third-hand, but at a debate held at the Oxford University Museum in 1860, Huxley became involved in a heated exchange in which Samuel Wilberforce, the Bishop of Oxford, asked him whether it was through his grandfather or his grandmother that he claimed his descent from a monkey. Huxley is said to have replied that he would not be ashamed to have a monkey for his ancestor, but he would be ashamed to be descended from someone who used his great gifts to obscure the truth. It sounds mild (even polite) now, but it was a powder keg under Victorian etiquette. Indeed, so outrageous was Huxley's counter-snipe and the constitution of the well-to-do so fragile that, in the turmoil that followed, one Lady Brewster fainted. It should be remembered that less than seventy years previously, John Hetherington, a haberdasher on the Strand in London, created enough of a sensation to topple four women in the same manner. Hetherington's crime, for which he was arraigned before the Lord Mayor, was wearing the world's first top hat. Officers of the crown at the hearing referred to the hat as 'a tall structure having a shiny lustre calculated to alarm timid people'.

Huxley's equally shocking performance was all the more notable because, despite his reputation as Darwin's bulldog, for many years the theory of natural selection was just that to him – a theory. One that, at the time, lacked any empirical evidence of its precise mechanism, very much, it could be argued, like Lovelock's Gaia hypothesis, which was inspired by and written atop Ehrenberg's heap of skeletons or, to put it another way, billions of coccoliths.

Huxley was posthumously awarded due credit for his work with these tiny creatures when the most numerous species of coccolith-bearing phytoplankton *Emiliania huxleyi* – scientists today refer to it as Ehux – was named after him. As it turns out, Ehux and its ancient ancestors are perhaps the most astonishing creatures on the earth. To find out why, we should take a moment to consider what trillions of coccoliths, along with the globigerinae, have left behind – chalk. And some tips on how to find it.

Downland in Britain is found on the hills shaped out of the Southern Chalk Formation, itself a part of a larger formation that stretches north to the Lincolnshire and Yorkshire Wolds and across the English Channel and North Sea where it lies under a large part of north-eastern France and Denmark. The formation's largest feature in England, the 300 square miles of Salisbury Plain, is the biggest area of chalk grassland in north-west Europe. The Hampshire and Marlborough Downs and in turn the North and South Downs radiate from Salisbury Plain as does a spur stretching south-west of Salisbury to form Cranborne Chase and most of mid-Dorset. A wide corridor of chalk runs north-east from the Marlborough Downs to form the Chiltern Hills and then the low hills of East Anglia.

A quick look at a map can serve well in finding some distinctive features of downland. A 1:50,000 Ordnance Survey Landranger map of the area is ideal, but even one of low detail, such as a road atlas, will do. Place names can provide good clues. Look not only for 'down' – a name that originates from the Saxon word *dun*, for hill, but also for place names that feature the element 'bourne' or 'borne' in them. A bourne is a stream or spring found in chalk and limestone landscapes. Where a line of villages running along the base of a hill is apparent, parallel to the ridgeway, look for a line of springs. These occur where a permeable rock lies over a non-permeable one, such as when chalk overlies clay. Water sinks to the bottom of the chalk bed and runs along the top of the clay until it emerges from the slope as a spring. The name given to the valley through which a bourne flows is a 'combe' or a 'dene'. Look also for place names starting with 'Winterbourne' – a winterbourne is a bourne or stream that only flows in winter. When winter rainfall raises the water table within the chalk it also effectively raises the spring line up the slope from the base of the chalk. When the water table drops again during the summer, the winterbourne dries up and leaves a dry valley behind.

Look for contours on the map that reveal the telltale presence of an escarpment. The contours would typically be tightly packed on one side of a hill's ridge and much further apart on the other. As we will see later, not all escarpments are found on chalk, however. They also readily form where other hard rocks – sandstone, limestone and igneous formations like sills for example – lie over less resistant ones. You can even spot the telltale signs of a chalk or limestone escarpment on a map without contours. Look for a line of villages spaced equidistantly on a road that runs roughly parallel to a stream and you may have found the spring line. A village that is located, with others, along the spring line is, unsurprisingly, called a spring line settlement. The parish boundaries of these spring line villages will often extend from the verdant, fertile river valleys up to the ridgeway and beyond to the dip slope, but be very narrow in the direction of adjacent villages – a feature sometimes referred to as a 'long parish'. What encouraged parishes to develop in this way was the availability of a wide range of agricultural resources from common land grazing on the escarpment through good quality arable farmland in the valleys and waterfowl and fishing at the water's edge. This was reflected also in the size of the farms before agriculture started to be run in a more corporate fashion – hill farms were anything up to 1,000 acres (400 hectares), while the farms in the more fertile valleys would rarely exceed 150 acres (60 hectares). As chalk is so porous, surface water doesn't occur naturally above the spring line so look out for dewponds. Modern dewponds, built for the benefit of grazing cattle, are made from concrete, but a few old clay ones still survive.

Keep an eye out also for white horses and other hill figures. Among these, the Uffington White Horse on the Berkshire Downs is ancient and has been dated at around 3,000 years old, but the rest appear to be medieval or Georgian in origin – indeed, the Osmington White Horse completed in 1808 features George III himself on horseback and was marked out to celebrate his patronage of Weymouth a few miles to the south.

Look for buildings and walls constructed from red brick and flint. Flint is formed as nodules – fine-grained quartz pebbles – mainly in chalk, but also other limestones. Unlike those other limestones, most chalk is too soft for use as a building material, but flint is exceptionally hard and weather-resistant.

On scarp slopes, look for terracettes – horizontal ribs spaced about a foot apart and arranged as tiny terraces on the steepest parts. They are caused by the gradual downhill slip of thin soils such as those that occur on chalk which is accelerated by the feet of grazing sheep who then end up using them as tracks.

Huxley wrote up some of his findings in an 1868 article for *Macmillan's Magazine* called 'On a Piece of Chalk', a skilful reconstruction of much of the geology of the British Isles extrapolated from a small chunk of East Anglian downland. By then, the rolling hills of southern England had long occupied a unique place in our patriotic awareness. Indeed, if there was a national poll then or now to find the landscape that articulated the concept of 'England' best, the Downs of the east and south-east of Britain would win – and it doesn't take a genius to see why. Downland is both green and pleasant after all and, as all students of William Blake and members of the WI will be keen to inform you – perhaps, even in song – England *is* a green and pleasant land. The lyric to 'Jerusalem' comes from the introduction to Blake's epic poem *Milton* and it is inconceivable that the line 'In England's green and pleasant land' was not a reference to, or at least inspired by, the downlands and coast of Sussex where he lived. Carried on a rousing, rolling anthem, it's hard to stop the mind's eye from watching a cinematic flyby over numinous low hills – a patchwork of fields dissected by cool, clear streams – but in a country blessed with such a wide variety of scenery in different shades of green and varying degrees of pleasantness, what is it about this particular landscape that marks it out as so English?

You could argue that the gentle undulation of the Downs calls to mind a kind of topographical aura of self-possession and composure, where their stillness and serenity turn out to be scenic shorthand for the stiff upper lip. Or on the other hand, you might point to that other face of downland, the indomitable white cliffs of Dover, which are more burdened with patriotic symbolism than the verses of 'Jerusalem' and 'Rule Britannia' rolled into one.

Then again, with sheep grazing on verdant turf around the odd ash or beech tree and a view only interrupted by scenic woods in the distance, you can't help noticing how much of the Downs feel like a rough-around-the-edges Arcadian paradise, a worn and frayed version of classic English parkland, of the kind engineered to order in the eighteenth century.

All of these impressions may have a bearing on how we feel about our Downs, but one aspect that seems incontrovertible is its rustic ambience. Gazing out over the gentle green contours, I always find my mind drawn to thoughts of earthy things – of ploughman's pickle, of tweed and twill and the title pages of musty books with woodcut illustrations of elms and rooks. The Downs are uncompromisingly rural and who is to say that it is not that alone that gives them their distinctively English appeal? With that thought, we are led to the conclusion that this landscape does not represent England so much as resemble most of England as it once was. To return to that national poll on the landscape, an appeal based on nostalgia with patriotism, as any political operator will tell you, can win almost any vote.

Despite the softness and friability of its chief rock, there is something adamantine and unyielding about downland. Its rural spirit seems persistent and can even endure where huge populations are just a few miles away. The South Downs, for instance, lie just to the north of mile after mile of unfettered and seemingly uncontrolled coastal development. From Bognor to Brighton and beyond, a stretch of coastline around 30 miles (48 kilometres) in length, there is barely a few miles of undeveloped shoreline. Regency terraces give way to Victorian streets, then 1930s avenues and post-war estates of semi-detached houses. All are now bound together in a matrix of twilight towers, retirement flats, nursing homes and the occasional second home dressed up in faux art deco stylings – on winter days its blank expanse of windows peer blindly out to sea like an abandoned lighthouse. The ribbon of development has every town belching out a suburb which involuntarily excretes another as part of some urban peristaltic impulse, but the urge to develop has been held in check along the coast by the South Downs which have retained their old country character while also bequeathing a backyard wilderness to the populations of the urban coastal strip. Away from the coast, the villages are small, few and far between and the hills themselves would be completely deserted were it not for walkers out on weekend rambles along the ancient ridgeways.

Those self-same walkers pass dozens of prehistoric remains – round and long barrow graves of tribal chieftains long forgotten, field systems and causewayed enclosures – that tell us that the Downs were not always deserted. Their very survival, however, shows us that the landscape has not been inhabited or even, in many cases, ploughed, since. The ridgeways themselves are the surviving fragments of an ancient network of paths and droving tracks in place long before the Romans imperiously started drawing straight lines over the landscape as part of their military and social conquest of Britain. The purpose of the ridge-ways, like the tracks in the valleys long lost or subsumed into the alignments of country roads and footpaths, was trade. Once the hilltops and ridges had been deserted for the more fertile valleys, it may be that they afforded safe passage across hostile country.

Despite their picture postcard appeal, downs can be challenging, even hostile, places to live and that's largely because, wherever they are, from the Dorset Downs to the Chilterns, from Cranborne Chase to the white cliffs of Dover, they share a common heritage which thwarts intense agricultural productivity. One stretch of downland, in the true sense of the word, is made of the same chalk as another and has been shaped by the same forces into the landscape we know today. The chalk is part of a single formation laid down over around 80 million years and is over half a mile deep in places. This period of geological time, the Cretaceous, is unique in that the vast majority of chalk deposits were laid down during this time – indeed, the Cretaceous takes its name from the Latin for chalk, *creta*. But what exactly is chalk?

We are all familiar with blackboard chalk, but it is not

the same stuff that the white cliffs of Dover are made from. Blackboard chalk is made from gypsum, the same mineral – calcium sulphate – that is used to make plasterboard and plaster of Paris (there is a substantial gypsum deposit at Montmartre in the French capital, which is where it gets its name from) as well as a component of Portland cement, all of which makes gypsum the Swiss Army knife of the construction industry. It is also used as a soil improver and a coagulant for tofu.

Real chalk is different; it is a porous rock, a very pure type of limestone formed almost entirely from the mineral calcite, a crystalline form of calcium carbonate (Huxley's carbonate of lime) which, as we have seen, is largely the product of billions of tiny organisms.

The coccoliths which, in some species, are continually shed throughout the lifetime of the organism, sank to the seabed during the Cretaceous period, which marks something of a watershed between the old earth and the new, between a world full of dinosaurs and a world that starts to look more like the planet that we inhabit today. The end of the Cretaceous, approximately 65 million years ago, is marked by a defining catastrophic occurrence that almost wiped out life on the planet altogether when an event of Hollywood disaster movie proportions (most suspect the impact of a very large meteorite) caused or was a major contributory factor in mass extinctions, including those of the dinosaurs. There is even a likely looking spot for the impact in the shape of the Chicxulub crater, the centre of which is located just off the coast of Yucatán in Mexico. The crater is 110 miles (176 kilometres) in diameter and, scientists believe, the result of a collision with a meteorite at least 6 miles (9.5 kilometres) wide. The impact of the

Chicxulub meteorite seems to have been the final straw for an ecosystem already under severe stress, stress caused by enormous releases of sulphur dioxide from an exceptionally large volcanic formation, the Deccan Traps in India, where as much as 1.5 million square kilometres of land was inundated with lava. This led to runaway global warming, habitat eradication and changes in rainfall patterns across the world.

Up until these stresses occurred in the last few million years of the Cretaceous period and before a rock the size of Liverpool dealt the fatal blow, there was extraordinary biodiversity and conditions for life were at the luxury end of hospitable. Most of the country lay under a warm and shallow sea in a climate often compared to that found in the modern Caribbean. In truth we have no modern equivalent to the Cretaceous seas. Even Huxley's ooze of globigerinae, so similar to the sediments that formed the chalk in the Cretaceous, are to be found today at the bottom of cold, deep oceans completely unlike the greenhouse marine habitat of the original chalk sediments. This is where the view of uniformitarianism – that the present is the key to the past, as first proposed by James Hutton – breaks down. It seems that chalk is the product of a singular chapter of history.

Everything about the Cretaceous can be expressed in hyperbole. Everything was bigger, more abundant or exaggerated in some way. Sea levels were higher – about a third of the world's land mass today was under water at the time. Atmospheric gases were off the scale: carbon dioxide concentrations were six times modern pre-industrial levels and oxygen made up 30 per cent of the atmosphere, almost half as much again as the present day. Average global temperatures were 5°C higher and the surface of the tropical sea may have briefly reached 42°C – more like a hot tub than an ocean. The earth was so warm that it is likely there were no polar ice caps, while there's also evidence to suggest that forests existed in Antarctic latitudes. Some palaeo-climatic scientists go even further and suggest that there was nothing short of a polar heatwave during Cretaceous summers, along with relatively mild winters. The presence of deciduous trees at latitudes of 82°N – well within the Arctic Circle and as far north as the most northerly bits of Greenland – during the middle of the period is further evidence that climatic conditions were very different from the modern world.

The Caribbean analogy does have a saving grace. If the thought of a Caribbean ocean calls to mind shafts of sunlight piercing a cerulean sea, we can at least imagine what standing on what is now Salisbury Plain was like between 100 million and 65 million years ago. We would have been standing in perhaps 200 to 300 metres of warm water some way away from land.[2] Around us the water may have had an unusual milky appearance on account of marine snow, the precipitation of millions of tiny smidgens, the calcite tests of micro-organisms on their way to the calcerous ooze below. We may have seen mosasaurs, not a dinosaur but a very successful air-breathing marine reptile from a family of predators. They ranged in size from 10 to 56 feet (3 to 17 metres) long and had double-

2. We know we would have been some distance from any land because chalk is made of the shells of tiny aquatic creatures and of the very few impurities found in it, none has an origin on land – sediments washed out into the sea from river estuaries don't tend to travel very far.

hinged jaws like modern snakes to enable them to swallow prey whole. Resembling a cross between a conger eel and a monitor lizard, mosasaurs gave birth to live young and prospered in the shallow Cretaceous seas. Mosasaur fossils have not yet been found in Britain but have been discovered in Belgium and the Netherlands among many other places, and during the last half of the Cretaceous much of Britain would have been covered by the same kind of shallow ocean – the Chalk Sea, to give it its geological name.

Despite their microscopic dimensions, coccolithophores are vital to the well-being of the environment and may even be instrumental in keeping the planet at an optimum temperature for life. What is known is that they extract carbon dioxide from the atmosphere by photosynthesis, building coccoliths from it which eventually drop to the sea floor (an estimated 1.5 million tons of calcium carbonate are deposited every year in this way), at which point the processes of geology eventually form stable rock from the sediments, all of which make these tiny organisms responsible for the largest carbon sink on the planet.

All of which is not lost on Dr James Lovelock, the author of *Gaia: A New Look at Life on Earth*, which hypothesises that the whole earth, both the organic and inorganic parts, including the rocks, participate in a complex web of chemical cycles that regulate conditions on the planet to a common advantage. Initially, Lovelock's contribution was subject to a certain amount of scientific ridicule in particular because, for ease of reading, he referred to the unknown engine of a self-regulating earth as 'she', something of a Mother Nature figure. Though he provided a disclaimer in the preface stating that this literary anthropomorphism was no more significant than a ship's captain calling his old tub 'she', it still seemed to put some scientists off the idea, because it was not written in a sufficiently humourless style. It is a little like turning up to a board meeting with a revolutionary idea that will save your company millions of pounds and not being heard because you failed to prepare a PowerPoint presentation on it. In a subsequent edition of the book, Lovelock wrote:

> Because of my ignorance twenty-six years ago, I wrote as a storyteller and gave poetry and myth their place along science . . .
>
> None of [the scientists] appeared to notice the disclaimer, nor did they read the ten or so papers on Gaia in peer-reviewed scientific journals. The critics took their science earnestly and to them mere association with myth and storytelling made it bad science. My disclaimer was about as much use as is the health warning on a packet of cigarettes to a nicotine addict.

Coccolithophores are singled out in particular by Lovelock and it appears that increasing amounts of atmospheric carbon dioxide have been accompanied by a sharp increase in algal blooms in the world's oceans. During these blooms, the numbers of cells of Ehux can cover an area of ocean the size of England and often outnumber those of all other species combined, accounting for 80 per cent or more of phytoplankton cells in the world's oceans. During such a bloom, the surface of up to 100,000 square kilometres of ocean will change colour to a milky turquoise, increasing its albedo and reflecting three times as much sunlight back into space.

Perhaps more importantly, when coccolithophores die

they produce dimethyl sulphide (DMS), a gas that smells like cabbages but helps build clouds, which cool the surface of the planet even more. Under conditions of human-influenced global warming it is uncertain which way this feedback loop will run. It will either contribute to homeostasis and help self-regulate the system or, in situations where the phytoplankton that produce DMS are negatively affected by one particular outcome of increased temperature, the effect will run away in the opposite direction.

Fifteen million years (not long, geologically speaking) after the last of the Cretaceous chalk of Bowerchalke Down, close to Lovelock's Wiltshire home, was deposited and the asteroid which would extinguish the dinosaurs struck, something even larger – though much slower – collided with Europe. This very slow something came to be known 50 million years or so later as 'Africa'.

The precise timeline of events is complex but, in essence, from the end of the Late Cretaceous to only 9 million years ago, the African continental plate crumpled up the south of Europe forming the Alps, a process known as the Alpine Orogeny. With such huge forces at work – the seabed was thrust to a height of something like 28,000 feet (8,534 metres) though the Alp's highest point, Mont Blanc, is only 15,774 feet (4,807 metres) today – the Alpine Orogeny was felt throughout Europe. In some parts of Dorset and the Isle of Wight the folding was so extreme that the beds of chalk are practically vertical. Along with the rest of south-east England, the South and North Downs are, in effect, what remains of the furthest foothills of the Alps, the outer ripples of a turbulent pond.

To make downland, those ripples – the gentle synclines and anticlines – had to be roughed up somewhat by

Formation of an escarpment

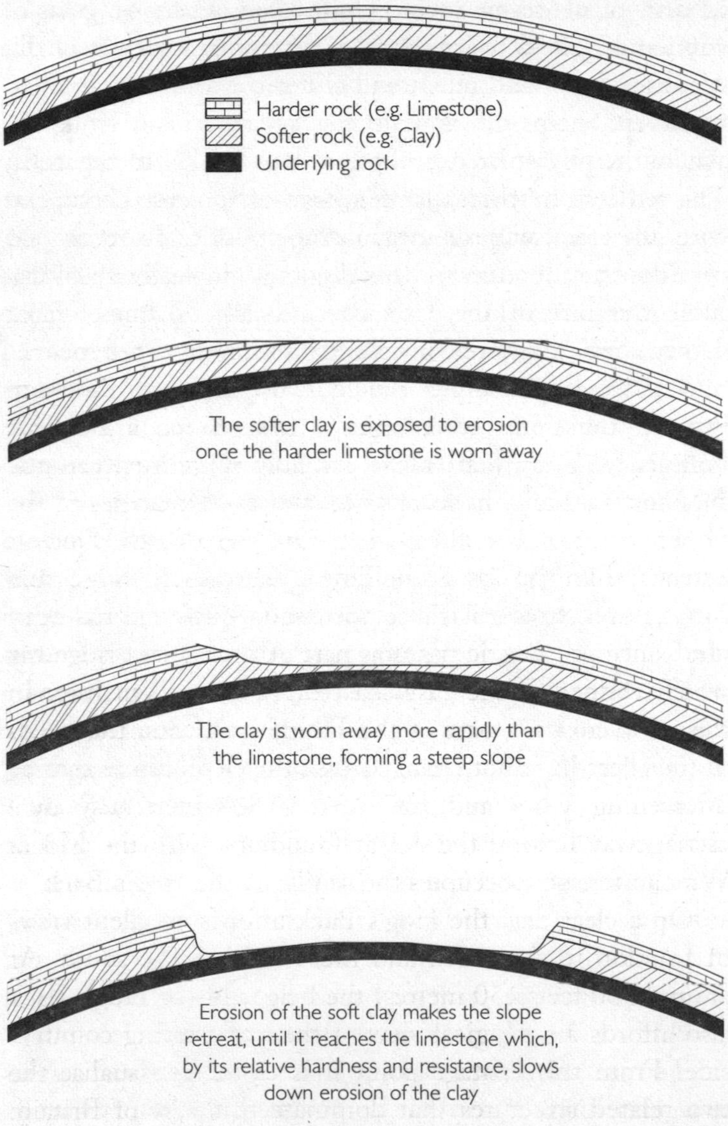
Harder rock (e.g. Limestone)
Softer rock (e.g. Clay)
Underlying rock
The softer clay is exposed to erosion once the harder limestone is worn away
The clay is worn away more rapidly than the limestone, forming a steep slope
Erosion of the soft clay makes the slope retreat, until it reaches the limestone which, by its relative hardness and resistance, slows down erosion of the clay

erosion. The magnificent escarpments of the North and South Downs we know today were fashioned by millions of years of whittling away by individual raindrops, gusts of wind and crystals of ice. Where softer rock underlay chalk, it eroded faster and retreated back into a steep slope called the scarp. Scarps run down across the rock beds, while the shallow angle of the dip slopes follow the tilt of the beds. This process continues to this day, not only in chalk, but with other limestones and sandstones which dip gently and overlie softer deposits. There is something special about a chalk escarpment however – a roundness, a softness – that makes a milder, more convex scarp face than other rocks.

The steeper the dip of the beds, the shallower the scarp is until the point of balance is reached in a hogback formation. Named after the knobbly ridge between the shoulders of a pig, hogbacks occur where the angles of the slopes are roughly the same. On the North Downs, between Farnham and Guildford, the narrow ridge that the A31 runs along is called the Hog's Back and has been used since prehistoric times as part of the ancient ridgeway which connected the east coast of Kent to Wiltshire. In this now crowded part of the Home Counties, the needs of travellers have not changed much over the thousands of intervening years and the A31, a furiously busy dual carriageway linking the A3 at Guildford with the M3 at Winchester, now occupies the saddle of the Hog's Back.

On a clear day, the Hog's Back affords excellent views of London to the north and the Weald to the south. At around 500 feet (150 metres) the height of the Hog's Back also affords a geological view of the surrounding country-side. From this vantage point it is easier to visualise the two related structures that dominate this part of Britain:

the London Basin[3] and the Weald–Artois anticline.[4] To the north, the London Basin has a structure that is reminiscent of a traditional dew pond – a dug-out depression in chalk with a clay lining. The chalk, of course, is not dug out but dips under the more recent London Clays to re-emerge north of the capital as the Chiltern Hills. The London Clay, meanwhile, a stiff, blue clay that lies over the chalk, has turned out to be an almost ideal medium on which to build a large relatively low-rise city. While constructing skyscrapers on it is problematic without extensive piling, it does make very good bricks. It has also proved to be an excellent medium for tunnelling through while it has long been recognised as being disastrous for agriculture. The development of the Tube was greatly aided by the stable yet soft rock underneath the capital while during the nineteenth century, in still-agricultural Middlesex, bringing the London Clay to the surface was known as 'ploughing up poison'.

Looking south from the Hog's Back over the Weald, you are gazing into the centre of a collapsed anticline, the crest of which may have been around 3,300 feet (1,000 metres) above sea level, an elevation somewhere between the current height of Wales' and England's highest moun-

---

3. A basin is a roughly circular syncline or depression of the sedimentary beds. Its appearance on a geological map is characterised by younger rock at the centre surrounded by concentric rings of older outcrops.

4. An anticline is an upfold of strata with the oldest rocks at its core. A simple aid to distinguishing between an anticline and its opposite – the syncline – is that an anticline 'points' up like the capital 'A'. Another, perhaps simpler, way is to think of a syncline as a 'sink', which also neatly ties in with the idea of a 'basin' (see previous footnote). The opposite of a basin is a dome, a roughly circular anticline.

tains, Snowdonia and Scafell Pike, respectively. The eroded-away centre of the anticline on which the Weald lies consists of sandstones and yet more clays, although the verdancy will tip you off that they don't plough up poison in these parts: this is not London Clay but the earlier Gault Clay deposited before the chalk in the Lower Cretaceous.[5]

Between the two escarpments of the North and South Downs, the Weald actually contains escarpments of its own. Just as chalk downs surround the Weald, a ridge of greensand – a rock deposited as sand in shallow marine environments – forms parallel concentric ridges to the south of the North and to the north of the South Downs, while the more heavily eroded softer clays lie under the valleys. It is almost a repeat of the same basic pattern we see on the Downs, except that the character of the landscape here is totally different. The core of the anticline, the High Weald, which occupies the eastern end and stretches from Hastings to Horsham, features fine-grained sandstone with high rocky outcrops standing over clay vales of variable quality for agriculture: some are essentially sterile, while others are rich.

The name Weald comes from a word for wood. According to the Domesday Book the High Weald was the most wooded area in England and you still get a sense of that nearly a thousand years later, even though there has been extensive deforestation. Trees were cut down for shipbuilding and for charcoal for a local iron industry that has since moved on to pastures new: South Wales received

5. Units of geological time can be divided into Upper, Middle and Lower. Lower is earlier, Upper equals later in the geological scale.

an influx of Wealden ironworkers when the industry faltered at the beginning of the nineteenth century. All that remains, aside from the occasional museum display, are the place names – Furnace Pond, Minepit Wood, Forge Lane and Hammer Mill being typical examples – and a number of excavations of sites that were worked as far back as the Roman invasion. The source of the iron was ironstone in the greensand and clays, while building stone and clay brick were in abundance to build kilns and furnaces and fast-flowing streams powered mills to pump bellows. It was only the adoption of coke smelting that put an end to the Wealden iron industry for good.

The collapsed anticline in which the Weald lies offers us a view through the chalk to what lies underneath. Most of it is Lower Cretaceous but between Battle and Heathfield all of the Cretaceous rocks have been eroded away leaving three small inliers,[6] three glimpses of a world even older: Jurassic rocks.

---

6. An inlier is an 'island' of older rock formation isolated in an area of newer rocks. An outlier is the exact opposite.

CHAPTER SEVEN

She Sells Sea Shells . . .

Jurassic period – from 146 million to 200 million years ago

By all accounts the life of Mary Anning, the subject of the famous tongue-twister *She sells sea shells by the seashore*, was not easy. She seems to have been cursed by circumstance while being blessed with perseverance in equal measure and her life is an extraordinary story of destitution, tenacity and, finally, recognition of her considerable talents as a fossil collector. In 1800, at the age of fifteen months, she was the only one of four people caught in a lightning strike in Lyme Regis who survived the experience and on another occasion, later in life, she almost drowned. This pair of near escapes inspired the lawyer John Robert Kenyon to some extremely bad verse, when he wrote:

> To Mary Anning:
> Thee, Mary! First 'twas lightning struck,
> And then a water-vat half drowned;
> But I can't think 'twas mere blind luck
> Twice left for dead – twice brought thee round.
> No! Fortune in her prescient mood,
> I must believe, e'en then was planning
> To fabricate a something good
> Of Thee, the twice-saved Mary Anning.

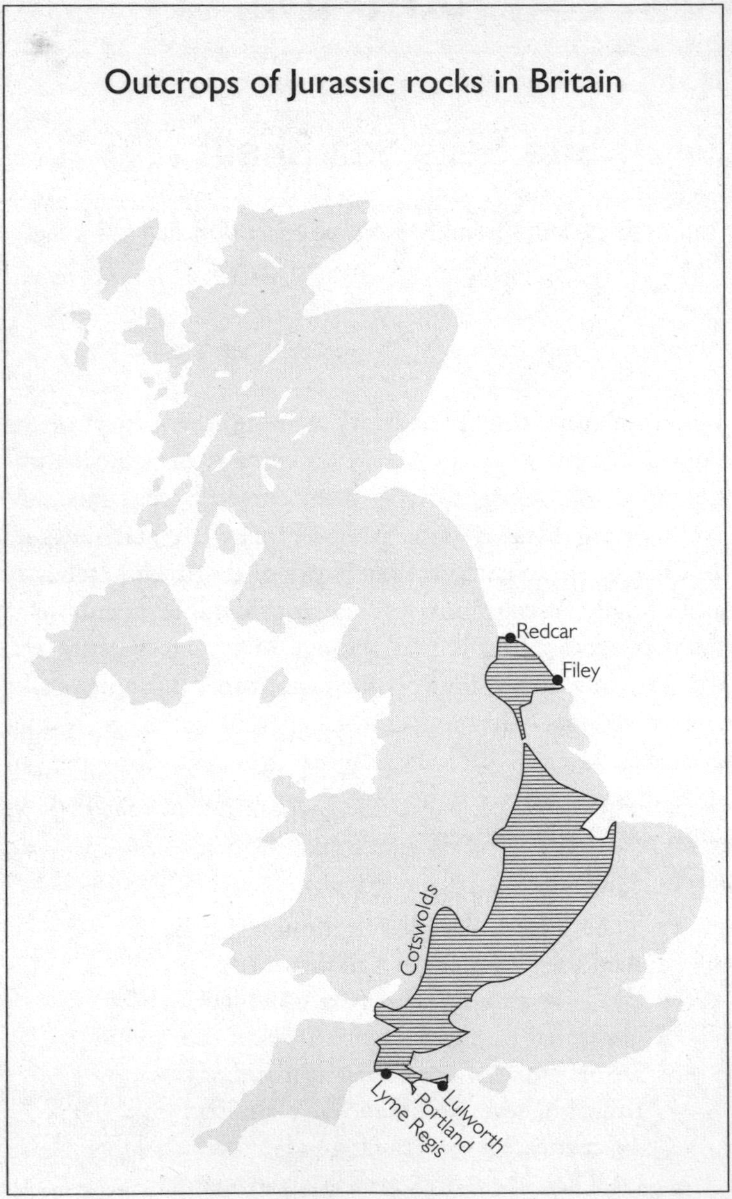
Outcrops of Jurassic rocks in Britain
Redcar
Filey
Cotswolds
Lyme Regis
Portland
Lulworth

Terrifying doggerel aside, between the two near misses, Mary lost her father, Richard, to a combination of tuberculosis and a bad fall when she was only ten years old. With the family left without any means of support, Mary and her brother Joseph turned to the beaches and cliffs of Lyme Regis to hunt for fossils and sell them to keep the family fed. For years their father, a cabinet maker by profession, had been a keen collector and the children had learnt where and how to find the fossils. Richard Anning had supplemented his income by selling his finds as curios and now his children could help support the family in exactly the same way.

By the age of eleven, guided by the find of a 'crocodile' skull by her brother Joseph, Mary Anning unearthed a complete fossilised skeleton of an ichthyosaur – a large marine reptile that looked, if reconstructions of it are anything to go by, a little like a bad-tempered dolphin. This tale is often simplified as being the first-ever complete ichthyosaur skeleton to be discovered, but it was, rather, the first one to come to the notice of professional scientific circles. Anning's ichthyosaur was around 18 feet (5.5 metres) long and it was soon described in the *Transactions of the Royal Society* by Sir Everard Home in 1814 in an article entitled 'Some Account of the Fossil Remains of an Animal'. Home's account lavished thanks on the owner of the fossil, who bought it for the princely sum of £23, but – as was often the case – the family who discovered and collected it were not even mentioned.

'Fossil collector' is an occupation that has the whiff of amateurism about it, but Mary Anning was more than a glorified beachcomber and merchant of curios even if that did become the chief trade of the Anning family. They would otherwise have had to rely on the parish poor relief. Away from the inequitable and unenlightened society of the nineteenth century, where everything was skewed in favour of gentlemen of means, this remarkable woman would have had a celebrated life for the finds she made – finds which helped to establish palaeontology as the science it is today. Having said that, if she had pursued the same interests two hundred years earlier, she might have wound up being burned at the stake. In retrospect, we can safely change her occupation from mere fossil collector to palaeontologist because, despite her humble background, Anning read widely on geology and the experts of the day were more than happy to correspond with her, drop into the shop to pick her brains or go on fossil- and geology-seeking trips with her. They knew a palaeontologist when they saw one.

Anning's fame gradually grew, despite her gender, and she continued to find extraordinary fossils along the Dorset coastline, including a near-complete plesiosaur and a complete skeleton of a pterosaur (known also as a pterodactyl). She also drew attention to the stones that were often associated with the abdominal areas of ichthyosaur fossils. The geologist and Dean of Westminster the Very Reverend Dr William Buckland, the man who first described the fossilised bones of a dinosaur,[1] investigated and concluded that the bezoar stones,[2] as he first called them, were

1. Buckland described the dinosaur megalosaurus, which was named the year after its discovery by Buckland as *Megalosaurus bucklandii* by Gideon Mantell.

2. Buckland called them bezoar stones, because they bore some similarity to stones taken from the guts or livers of goats and deer. As any student of either medieval medicine or Harry Potter will tell you, bezoar stones were alleged to be a universal antidote to all poisons.

fossilised faeces. Buckland even inlaid a collection of them into an ornate side table which, according to his son, was much admired by people who didn't have the faintest inkling that they were looking at a presentation of fossilised reptile crap. In 1829, Buckland published his conclusions at Oxford, where he was the university's first Professor of Geology. A rhyme of the times, popular around Oxford, immortalised the important discovery.

> Approach, approach, ingenuous youth
> And learn this fundamental truth:
> The noble science of Geology
> Is bottomed firmly in Coprology.

Anning finally received as much recognition as the times allowed. She was given an annuity by the British Association for the Advancement of Science and made an honorary member of the Geological Society – as a woman, she was not eligible for normal membership. When she died at forty-seven of breast cancer, members of the Geological Society made contributions towards a stained-glass window in Lyme Regis church.

Like most of us, measured against the Annings, my 'fossilising' is not going to lead to a pension from the British Association any time soon, not even in my dotage. Nevertheless, I wanted to see exactly how easy it might be to find and take home that iconic fossil, the ammonite, which is the star turn of the fossil world. Wherever you see the word 'fossil', there is likely to be a drawing of one very close by; there are even two representations of an ammonite integrated into the cover of this book. Although now extinct, ammonites were a group of cephalopods whose nearest modern relatives are octopuses and cuttlefish, except that they retained a spiral shell. The agreed interpretation of what a live ammonite may have looked like is a little like a squid stuffed into a snail shell. All of which makes the ribbed, usually planispiral[3] shells of ammonites the fossils that we all instinctively know the name of. They are also remarkably common, especially in the Jurassic period where they evolved rapidly and left behind a wide range of species that can be used with some precision to date the rocks. The area around Lyme and Charmouth is rich in ammonites.

A wander along Monmouth Beach, to the west of the spectacular thirteenth-century harbour wall known as the Cobb, will take you to the 'ammonite graveyard'. While the site is protected and you are not allowed to hammer out or collect these fossils, standing on the flat rock of the foreshore of Monmouth Bay within sight of hundreds of fossils is an eye-opening and rather humbling experience – some of the ammonites are 3 feet (1 metre) across. The fossils are embedded in strata known as thc Blue Lias, a sequence of early Jurassic limestone and shale layers between 195 million and 200 million years old.[4] Blue Lias is a building stone, but also is important as a source of lime for mortar – where its high clay content (for a limestone) makes it suitable to be used in hydraulic lime, a mortar that can even set underwater.

The Blue Lias itself was set underwater and, aside from the odd dinosaur corpse caught in a flash flood or the pterodactyl that presumably dropped out of the sky and

3. There were a few helically spiralled ammonites and some without a spiral at all.

4. If you say 'layers' in what you might imagine to be a Dorset accent, you will find where the name lias came from.

into the sea, there aren't any fossils in the Blue Lias of the kind that starred in *Jurassic Park*. But, as a glance at Monmouth Beach will show you, as well as a similar area of flat rock known as the Broad Ledge on the east side of Lyme, there are fossils aplenty of Jurassic marine fauna.

Lyme and Charmouth, a few miles to the east, are at the centre of the Jurassic Coast, a 95 mile (153 kilometre) stretch of cliffs, coves and seaside famed for its extraordinary geology. The coastline, which is a UNESCO World Heritage Site, is not wholly Jurassic, but follows the geological succession from the Triassic New Red Sandstone at Exmouth to the Cretaceous chalk of Old Harry Rocks, south of Poole Harbour, near the village of Studland. Because of a gentle tilt of the rocks, by travelling from east to west, you are effectively travelling back in time – a total of 185 million years for the whole coast. The variety of landscapes on offer along this coast is breathtaking, but if you are thinking of collecting fossils for the first time you should head to Charmouth, where you will find the Jurassic Coast Heritage Centre and a number of ever-so-obliging outcrops and ammonites that seem on occasions to rain down from the rapidly dissolving Blue Lias cliffs.[5] The best time to go is on the falling tide, where fossils that have been washed out from the cliffs get caught in rockpools along with their modern marine counterparts.

I defy anyone with an hour or so to spare and a

5. I'm not particularly a fan of the Fun Police, the forces that seek to rule out all risk, no matter how insignificant or trivial, and extinguish all adventure, but around the Lyme and Charmouth areas the risks are not trifling and I urge you to take care and to never climb the cliffs – they really are unpredictable and chunks often fall off after inclement weather or a rough tide. Keep an eye on the tides and be aware that the area is prone to landslips.

moderately keen eye to *not find* something they can take home on this beach. I was in just such a position on a family jaunt to the seaside. I had forty-five minutes or so free after a picnic on the beach, during which the rest of the clan were occupied paddling around and sandcastling to their heart's content while I went off for a walk up the beach. By the time I had returned, I had a piece of shale with half a dozen perfectly preserved ammonites in it, all of which were between a centimetre and an inch in diameter. As I walked back, I passed a small party of guided walkers, one of whom had discovered a delightful little ammonite preserved in iron pyrites, a mineral often known as 'fool's gold'. If you have a bit more time to spare, these walks are excellent value and can take you directly to the most interesting spots.

Fossils aside, this part of the Jurassic Coast is well known for its landslides. The most notable landscape to be created by one is the Undercliffs National Nature Reserve between Seaton and Lyme, a landscape immortalised in the book and film *The French Lieutenant's Woman*. This is such a young landscape that parts of it were created within living memory. One major landslide on Christmas Eve 1839 at Bindon, a few miles east of Seaton, attracted thousands of visitors – some even came by paddle steamer to get a glimpse of the enormous chasm created when a 500-metre-wide chunk of clifftop detached itself from the rest of East Devon and slid as a block towards the sea. That block is now known as Goat Island, although it hasn't reached the sea so far. I used to walk up on the landslip as a teenager, in a place called, locally at least, the Elephant's Graveyard a couple of miles to the east of Seaton. My friend and I used to collect golf balls hopelessly sliced off the tee by members of the Axe Cliff Golf Club and sell them to anyone who wanted to take advantage of these fruits of the higher handicaps. I notice now that the course is somewhat smaller than I remember, so maybe that's one of the positive pay-offs of coastal erosion – the exposure of fossils and the slow but eventual elimination of golf. Win-win.

If all the golf clubs in the world sank into the beautiful wildernesses of the sort found at the Undercliff, I don't think even golfers would complain. It is the nearest thing to a verdant wild wood in these islands. It has an enormous – the biggest in Britain – self-sown ash wood, the product of a sheltered and rather humid microclimate that hosts many different habitats from scrub to dense woodland. Even if it wasn't a national nature reserve, its shifting strata means that the likelihood of it being developed is somewhat less than a Tokyo skyscraper being built on foundations made of cream crackers. What is even better is that the whole Undercliff is on the South West Coast Path, a challenging and strenuous 7 mile (11 kilometre) walk which you have to commit to: the only way out is to carry on to the end or retrace your steps.[6]

Although chaotic and complicated, a landslide is, like many phenomena in the landscape, an instance of forces attempting to reach some kind of equilibrium. Like all landslide-prone landscapes, this area is particularly susceptible because of the underlying geology. Here, porous rocks like the chalk and the greensand from the Cretaceous lie over impermeable mudstones and the Blue Lias. When the greensand or chalk become waterlogged after heavy,

---

6. You shouldn't leave the path either as, apparently, there are deep fissures hidden in the verdant vegetation.

prolonged rain the rock becomes much heavier as a consequence. The junction between the porous and impermeable strata also becomes lubricated and the top block can slump off the cliff in large pieces, effectively rotating and sliding along the slip plane making the top tip back towards the cliff, in a movement similar to a cartoon character slipping on a banana skin. The slope is always trying to move towards equilibrium, but the sea continually removes the foot of the material that has slipped down to the shore and the rock's natural angle of repose is never achieved for very long.

Further east from Charmouth is Golden Cap which, at 626 feet (191 metres), is the highest point on the south coast of England. In common with the other cliffs along this stretch of coast, early Cretaceous greensand forms the porous and unstable peak teetering on the Middle and Lower Lias clays. Although you should perhaps not linger especially long on the narrow beach under Golden Cap, you should try to steal a look at the great cliff and remind yourself that all those layers took the best part of 80 million years to accumulate. Right at the very top, however, is the crowning glory of the south coast which, although it is formed from greensand, is distinctly golden yellow in colour. The cliffs have suddenly become much more vegetated in the last few decades and nobody really knows why, but Golden Cap must have been even more spectacular in the early half of the twentieth century when F.J. Harvey Darton visited, as recounted in his 1922 book *The Marches of Wessex, a Chronicle of England.*

> There is one other place in Dorset where the Earth's own past obtrudes itself, in a great view, upon one's

> thoughts about man's past and present. That is the summit of the highest cliff between the Wash and Land's End, Golden Cap. That glorious hill is known and loved by all Dorset men. It stands up with a peculiar boldness: a piled-up sloping mass, and then a bare stretch of yellow earth, crowned with a dark brown plateau.

If Golden Cap is iconic to the people of Dorset, then Chesil Beach is perhaps known to everyone else. Starting at West Bay, the coastline starts to describe a magnificent, almost geometrical, slightly curved line. Chesil Beach – or Bank – is an 18-mile (29-kilometre) shingle barrier beach which is slowly advancing on the coastline. There is no doubt that Chesil Beach is a striking feature – the view of it from Abbotsbury, roughly halfway along, was voted a rather begrudging 'Britain's third best view' by *Country Life* readers, who were probably all parked in unfashionably large 4x4s in a lay-by above the village, rattling their pearls and negotiating dowries for their unfeasibly ugly children. For my money, there's a better view, though, from the viewing area on the Isle of Portland at the eastern end of the bank. That view takes in the ferry port of Weymouth and the Purbeck coast to the east as well as the long sweep of Chesil.

It is often believed that Chesil Beach was formed purely as the result of longshore drift, where the waves drive stones up the beach at an angle, which then fall back under the force of gravity down the beach at right angles to it. Over the years, longshore drift moves the material on beaches along the coast in this sawtooth pattern until they reach an obstruction such as a headland or one of the breakwaters or groynes you might see on a seaside holiday. In this case the Isle of Portland acts as a natural groyne to stop the beach from moving any further east and, indeed, there are no Chesil pebbles around the corner in Weymouth Bay. However, longshore drift falls short of explaining the original formation of the bank as a whole.

There is no clear consensus on the exact mechanism, but Chesil Beach is believed to have formed as a result of a rise in sea level following the end of the last glaciation. It was predominately made of sand and gravel deposited further out in Lyme Bay around 120,000 years ago, and then driven onshore by rapidly rising sea levels between 10,000 and 5,000 years ago. Meanwhile, sea levels were low during the last glaciation and the landslides of the temporarily landlocked cliffs of East Devon and West Dorset continued, building up great piles of chalk and greensand across the sea floor. When sea levels started to rise again, huge quantities of flint and chert pebbles were washed out of the chalk and greensand where they had originally occurred as nodules. These pebbles were then transported along the coast by longshore drift.

Regardless of how it was formed, Chesil Beach is the largest example of a tombolo[7] in Britain and behind it, for half its length, is the Fleet – the largest tidal lagoon in these islands. The Fleet is an internationally important habitat in its own right, but Chesil Beach protects more than the wildlife of the lagoon; given the orientation of the coast behind Chesil Beach – which faces an unbroken fetch

7. A tombolo is a spit or bar of sand or gravel that connects an island to the mainland, often formed as a result of the refraction of tidal currents around the island.

of waves that stretches the whole way across the Atlantic Ocean – it effectively protects the port town of Weymouth from regular flooding and severe erosion.

Beyond Weymouth, to its end at Old Harry Rocks north of Swanage, the Jurassic Coast shows both of the two classic types of coastline, concordant and discordant, and all of the features that may form on them. When taken with Chesil Beach, the landslides and the ria of Poole Harbour, the Jurassic Coast is the next best thing to a coastal pattern catalogue, with fine examples of many features all within the same 95 miles of coastline.

A discordant coastline, such as the one found on the east coast of the Isle of Purbeck between Durlston Head and Poole Harbour, is where a succession of beds of different rock types outcrop, as they do when the bands of rock run perpendicular to the shore. Jurassic Purbeck Beds of resistant limestone form the headland at Durlston Head, while the Foreland or Handfast Point to the north is the seaward finger of the high, narrow ridge of the Purbeck Hills which features at its crest pure white chalk from the Cretaceous period. Between them the parabolic sweep described by Swanage Bay owes its embayment to the relatively soft Wealden Clay. North of Handfast Point, the erosion of Eocene clay and sand are responsible for Studland Bay.

The south coast of Purbeck from Durlston Head to Weymouth Bay is just as superb an example of a concordant coastline, one where the bands of rock all run parallel to the shore itself and just one type faces the sea. The bands are narrow and the beds dip steeply, so at Lulworth Cove the hard Portland stone and Purbeck limestones forming the face of the cliff are backed by Wealden Clays within metres, which are then succeeded by the chalk a short distance to the north – less than 400 feet (130 metres). This rapid succession of rocks of varying characteristics is what gives Lulworth Cove its shape. A stream originally breached the Portland stone where the mouth of the cove is today, a narrow gap in this limestone was then exploited by the sea. Once it had worked its way through to the much softer Wealden Clay, it rapidly eroded out the circle of the Cove until it reached the harder chalk forming its northern wall. A few hundred yards to the west of Lulworth Cove is Stair Hole, a snapshot of a cove like Lulworth, in its early stages of formation, but also the site of the magnificent twirling bedding planes that are known as the Lulworth Crumple. The strata of the Crumple were buckled by the same tectonic forces that raised the Alps.

Today Lulworth Cove is probably the most popular location in Britain for geology and geography field trips, as well as day-trippers. Its popularity almost overshadows the other wildly popular scenic spots on the same coast. A mile and a half west of Lulworth is Durdle Door, a perfect natural sea-arch, while to the east lie the remains of a fossilised forest, from the late Jurassic, when sea levels subsided leaving a number of small islands across Dorset under the canopy of giant cypress trees, each island surrounded by salt lagoons.

While there are patches of Jurassic rocks as far north as the borders of Caithness in Scotland, away from the south coast the most significant outcrop of the limestones, sandstones and clays of the period form a broad sweep of hills that run from Bath in Somerset and into the Midlands: the Cotswolds. Here, virtually every house is built in the warm sepia-toned local oolitic[8] limestone – Cotswold stone. Villages such as Broadway in Worcestershire encapsulate what many of us think of as the perfect expression of the picturesque village. A wide street (indeed, a broad way) runs through the centre of a postcard village with honey-coloured cottages and well-clipped green verges. A few miles to the south, the village of Stanton was described by Nikolaus Pevsner as 'architecturally, the most distinguished of the smaller villages in the North Cotswolds'.

That might sound like faint praise, but for Pevsner to even mention 'architecture' and 'distinguished' in the same sentence, required a critical leap. In fact, he once wrote that: 'A bicycle shed is a building; Lincoln Cathedral is a piece of architecture. Nearly everything that encloses space on a scale sufficient for a human being to move in is a building; the term architecture applies only to buildings designed with a view to aesthetic appeal.'[9]

Without a doubt, the aesthetic appeal of the Cotswolds stems mainly from the stone used to such great effect in the buildings and drystone walls. Indigenous building stone is always sympathetic to the landscape that provides it, but here there is so much agreement between land and architecture that the occasional modern brick building is almost considered an affront. In *An English Journey*, J.B. Priestley wrote of the Cotswold stone and the character and charm of the walls that are built from it: '. . . the truth is that it

8. Oolitic limestone is inorganic, i.e. formed from chemically rather than biologically derived sediment like coral or shells. It is composed of tiny egg-shaped spheres, or ooliths, of calcium carbonate. They precipitate from warm, shallow calcium carbonate-rich water where there are strong currents.

9. *An Outline of European Architecture*, 1943.

has no colour that can be described. Even when the sun is obscured and the light is cold, these walls are still faintly warm and luminous, as if they knew the trick of keeping the lost sunlight of centuries glimmering about them.'

Both Stanton and Broadway lie to the north-west of the magnificent Cotswold Escarpment. It is said that on a clear day you can see up to thirteen counties from the top of Broadway Tower, a 55 feet (17 metre) tall folly perched on the top of Broadway Fish Hill – itself the second highest peak in the Cotswolds at 1,024 feet (312 metres) above sea level.

The A44, which passes close to the Tower, crosses the map in long straight green diagonal strokes on its way from Evesham south-east to Oxford, all except for the section where it climbs the Cotswold Escarpment at Fish Hill. It does so in a series of twisting bends which look as contorted and sinuous on the pages of a road atlas as a short stretch of small intestine. The view from the ground, from the Fish Hill picnic area, is stunning, but that is a relatively modern perspective. In the days before picnic areas and lay-bys, before the signs that obligingly remind you to 'try your brakes', before 'escape lanes' and safety barriers, another road laboured up past the now-abandoned quarry, recently re-purposed as an amenity area for the consumption of neat sandwiches in Tupperware boxes. The 'Old London Road' would have once resounded with the clatter of hoof and coach and cartwheel on stone – extra horses were frequently taken on in Broadway so coaches could tackle the escarpment – and once on the top of the Cotswold Edge, what then?

Our ancestors had an uneasy relationship with all upland right up to the nineteenth century. Even those on the Grand Tour in Europe would hurry through the Alps as if it were populated by barbarians and unknown monsters. Where now we see wilderness, space and freedom, they saw only menace, austerity and gloom, so they regarded a view like the one from even the modest upland of Fish Hill as a blessed relief. Sydney Smith, the clergyman and writer, summed up the consensus view of the eighteenth century in an account he gave of the sudden appearance of a view from the escarpment when he was travelling over the edge, a little to the south:

> The sudden variation from the hill country of Gloucestershire to the Vale of Severn, as observed from Birdlip, or Frowcester Hill, is strikingly sublime. You travel for twenty or five-and-twenty miles over one of the most unfortunate, desolate countries under heaven, divided by stone walls, and abandoned to screaming kites and larcenous crows; after travelling really twenty, and to appearance ninety miles, over this region of stone and sorrow, life begins to be a burden, and you wish to perish. At the very moment when you are taking this melancholy view of human affairs, and hating the postilion, and blaming the horses, there bursts upon your view, with all its towers, forests, and streams, the deep and shaded Vale of Severn. Sterility and nakedness are thrown in the background: as far as the eye can reach, all is comfort, opulence, product, and beauty . . .

It is hard to reconstruct mentally Smith's Cotswolds from what most travellers see at Birdlip now, and not just because our ideas of landscape and the sublime have changed so thoroughly since his day. Once at the junction of an ancient ridgeway and Ermin Street, the escarpment proved too much for the wagons and carts which were increasingly in use by Smith's day and a turnpike took traffic away from the crossroads at the centre of the village. Birdlip is no longer the meeting point it once was; the turnpike turned into the A417, which is now a strategic link road to and from the M5. But the view west from Barrow Wake over Gloucester is still there and magnificent, though the dual carriageway of the A417 makes an unwelcome laceration across the landscape.

The crest of this hill belongs to the Birdlip Limestone Formation of the Middle Jurassic, one of a number of limestones that run the length of the Cotswolds known as the Inferior Oolite. Older and therefore lying under the Great Oolite – of which Bath stone in particular is among the world's finest building materials[10] – William Smith originally called it the 'Under Oolite'. What we know as the familiar Cotswold limestone often comes from the Birdlip Formation, which extends through Broadway and further north-east. If you find yourself on the Birdlip, as I did once, during early autumn when fresh ploughing has exposed the earth, you may notice the slightly odd colour of the fields. They are a peculiar orangey-brown rather like the aspirational 'gold' brands of instant coffee granules we all used to buy when our aspirations were much more modest. The colour also has an element of raw sienna about it – the colour of both the fields and of the vernacular architecture originates in the oxidisation of iron in the limestone to form a mineral called goethite and a secondary

10. Bath Stone was not only used extensively in high-status developments in Bath but also in many of the Oxford Colleges, Bristol Cathedral and Lancaster House in London.

mineral[11] called limonite. In Mediterraean lands, *terra rossa*[12] is the old name for red soils formed in much the same way, but limonite is perhaps better known as the pigment called yellow ochre. The brown ochre of our coffee-stained fields, meanwhile, is goethite.

Geological history is full of happy accidents like the formation of ochres, where a trail of unconnected chemical reactions over time has resulted in a mineral of great benefit to humanity millions of years later. Some of the fruits of these geological flukes have become so ubiquitous and familiar, their widespread exploitation has turned them from happy accident to potential calamity. The Upper Jurassic's limestones have an even more important deposit sandwiched amongst them. Kimmeridge Clay is named after the bay on the south coast of the Isle of Purbeck where there is a substantial outcrop. It doesn't sound that important until you realise that it is the major source rock for North Sea oil.

To the west of Kimmeridge Bay there is an oil well that is the oldest continually operational 'nodding donkey' in the world. Crude oil is formed in a similar way to coal, in that it is the product of anaerobic decay of organic matter but, like coal, a very specific and rare combination of circumstances have to occur in order to create an exploitable oilfield. Layers of mud quickly covered layers of dead plants and animals and stopped them from decomposing in the normal way. As layers of mud built up, exerting pressure on the organic matter, and the organic matter was buried further into the earth it was also heated

---

11. A secondary mineral is one formed from other minerals.

12. UNESCO, however, have wiped all Latin romance from them by calling the *terra rossa* earths 'chromic luvisols'.

by geothermal processes. Micro-organisms which can survive without oxygen – anaerobic bacteria – slowly broke down the matter, aided by heat and pressure, to form oil. The higher the temperature and pressure, the higher the quality of oil. (North Sea oil is high-quality 'light' oil.) This oil then migrates upwards through a porous sandstone that has been folded to form an anticline. The sandstone will contain water, but as oil is lighter, it will float up to the crest of the anticline. If the sandstone happens to lie under a rock that can form a seal over it, known as the 'cap', the oil will be trapped there in a reservoir.

As well as oil, the subject of building stone is never far away from a discussion of Jurassic landscapes, but the use of its limestones is not just a story of the Great and Inferior Oolites of the Cotswolds because the Jurassic saved the very best – arguably the world's best – building material till last: Portland stone.

Like many rocks that carry a regional identity, Portland stone does not just occur on Portland, though this is where it is quarried in huge quantities. Considering its role in the reconstruction of London following the Great Fire of 1666, its ubiquity in great city architecture and its proximity to the sea from where it could be transported with relative ease, it is surprising that it has only been used extensively outside its source area since the seventeenth century. Even though it was not the first building in London to be built from Portland stone, it was the construction of St Paul's Cathedral, with the use of the best part of 1 million cubic feet (29,000 cubic metres) of the stone, that cemented its reputation. It was later used for the British Museum, the Bank of England, the National Gallery, Admiralty Arch, Buckingham Palace and the

Mansion House in London, among many others, while in 1950, the United Nations Building in New York also used it. Portland stone has become as much a part of the landscape of our cities as the London plane tree. Which is a happy coincidence because, compared with other limestones, Portland – like the plane – seems to be particularly resistant to modern pollution.

Its durability and perceived permanence have also given it strong associations with memorials of human suffering. All Commonwealth War Grave Commission graves from the First and Second World Wars are made to a standard design using Portland stone, as is the Cenotaph on Whitehall. The Monument to the Great Fire of London is built from it and many city centres that were ravaged by the Luftwaffe, Plymouth and Coventry among them, were reconstructed in the modernist style, but behind facades of Portland stone.

Portland stone is so favoured because of the relative ease with which it can be quarried and dressed. It is a freestone, meaning that it can be cut in any direction and that it is relatively easy to work and has a uniform texture. It was formed in the last age of the Jurassic, between 150 million and 146 million years ago when the whole of Europe was an archipelago in a shallow tropical sea. In other parts of Europe similar, fine-grained limestones were being formed. One in Bavaria, Solnhofen limestone, was so fine that it was used as the original lithographic[13] printing plate. Solnhofen limestone had lain relatively untroubled by much quarrying, until the invention of litho

13. Although lithographic printing today uses metal plates, the first plates were of lithographic limestone, as the name implies.

printing, which caused quarrying to increase exponentially, at which point Solnhofen became famous for a quite different reason – fossils. The limestone was a Lagerstätte[14] which preserved soft parts of many fossils in incredible detail including, discovered in 1861, the feathers of a winged bird, the archaeopteryx or, to give it its German name, *Urvogel*: the original bird. In the fourth edition of *On the Origin of Species*, Darwin seized on the discovery:

> Until quite recently these authors might have maintained, and some have maintained, that the whole class of birds came suddenly into existence during the eocene period; but now we know . . . that a bird certainly lived during the deposition of the upper greensand; and still more recently, that strange bird, the *Archeopteryx*, with a long lizard-like tail, bearing a pair of feathers on each joint, and with its wings furnished with two free claws, has been discovered in the oolitic slates of Solenhofen. Hardly any recent discovery shows more forcibly than this how little we as yet know of the former inhabitants of the world.

In Portland stone, meanwhile, there are giant ammonite fossils up to 2 feet (61 centimetres) in diameter which have the helpful, if slightly over-cooked, Latin name of *Titanites giganteus*. Quarrymen on Portland believed them to be sea-serpents and called them 'conger eels' and a visit to Portland or the south of Purbeck will usually be rewarded with the sighting of one, as they often end up in walls built from the Portland Jurassic rocks.

---

14. A German term for a sedimentary bed rich in fossils. Its literal German translation is 'place of storage'.

CHAPTER EIGHT

The Age of Minerals

Permian and Triassic periods – from 200 million to 299 million years ago

If you have ever skimmed through a book on rocks, have an extraordinary memory for geography lessons or have been unwittingly triggered into a suggestible hypnotic state by an Open University geology programme at three in the morning, you may have heard of the Old Red Sandstone. Other than under those special circumstances, it might have passed you by and you will also be blissfully unaware of its younger cousin named, with a crushing inevitability, the New Red Sandstone.

We will come to the Old Red later in Chapter 11 but for now all you need to know is that they are related and that the red of the Old and New Red Sandstones comes from a common source, iron oxide, the same compound that forms rust. Iron oxide in sandstone tells us something important about the conditions of its formation – that it had a general absence of vegetation. This links all the various reds; on early land masses where the Torridonian sediments settled and where the sun, unfiltered by an ozone layer, was inimical to all forms of terrestrial life; in the blazing Kalaharian heat of the Old Red continent; and here in the Permian, which

The minerals and moors of south-west England

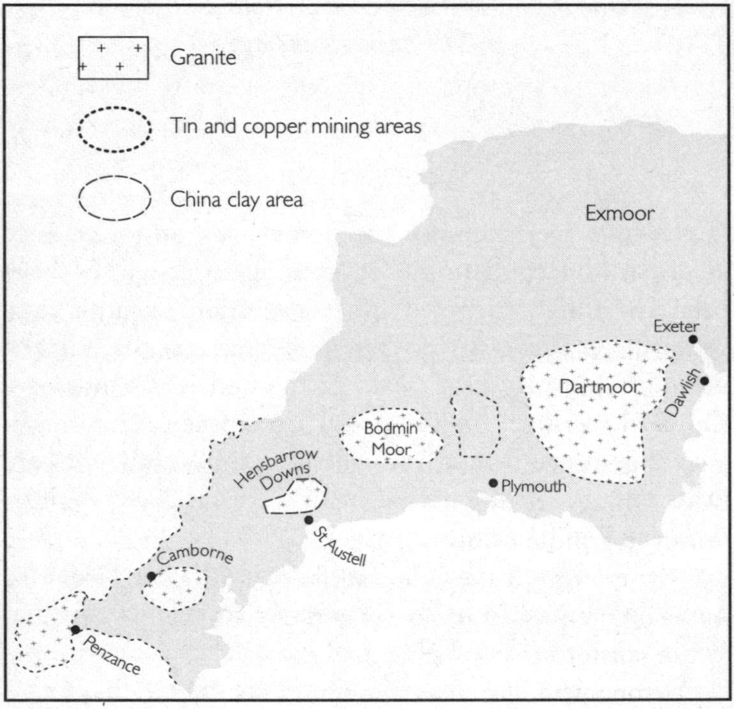

was basically the Old Red continent revisited – at the centre of a supercontinent of dust and salt – Pangaea.

Pangaea was the last of the supercontinents to have been formed in the geological record and, in common with all the others, it was brought together as the result of continental collision between the land mass we now know as Africa and the Old Red continent of Euramerica. This was the Variscan Orogeny. The Variscan – which in continental Europe, confusingly, goes by two other names, the Armorican or Hercynian – did not create any mountains in Britain, but shows itself in other ways.

Without even reading a word on the subject, the New Red's name suggests that it is a younger system of rocks than the Old Red. It was deposited over large areas of England in the Permian and Triassic periods and makes for fertile soils that inherit their striking colour from the rock. In some areas the New Red can be made from the erosion of nearby Old Red – the earth continually re-inventing itself anew as if run along Buddhist lines. But in Devon things are different; just as the Old Red was derived from the grinding down of the Caledonian, the New Red had the erosion of the Variscan mountains to the south as well as more local features created early on in the orogeny to the north to thank for its materials.

Anyone who has taken the train down to Cornwall will be familiar with the products of that erosion. After a glorious trip along the west bank of the wide Exe estuary where exotic waders like the avocet overwinter and where you might catch sight of an osprey on its migration to Africa, the line follows the coast through Dawlish Warren towards Teignmouth.

At Dawlish, the train passes alongside natural arches and

through tunnels cut into the New Red. These rocks, even on a dull day, are so bright, so energetic in colour, they seem to glow carmine[1] like the late afternoon light of a dusty day at harvest time. Yet closer inspection of them which is sometimes afforded by the delays that make Britain's rail system the warehouse of fun that it is, reveal intricate swashes and swoops in the bedding planes. These are fantastic examples of cross-bedding, each one the trace of a fossilised dune.[2] The grains of the sandstone are smooth and small, indicating that the forces of erosion, particularly the wind, have been working on every particle. Upon this ancient wind-blown sand, the effects of modern erosion, both wind and wave, can also be seen, sculpting out little natural niches into the cliffs, tearing out grains to accumulate again at some point in the future, perhaps creating what will inevitably be known as the Even Newer Red Sandstone. Further east, along the coast at Sidmouth, the New Red reaches the apotheosis of its expression along a discordant coastline of towering cliffs coloured, like those of Dawlish, as if bathed in evening light, only that of a deeper, later pink.

---

1. There is some speculation that the very name of Dawlish is a corruption of a Brythonic name – Deawlisc – which is taken from a local stream and means 'Devil Water'. After heavy rainfall, runs the explanation, the stream churns the cliffs and makes the stream run red. All of this, of course, sounds far too conveniently plausible to be true.

2. There is also pedestrian access via the South West Coast Path which runs along the side of the track, so you don't have to keep going back and forth, getting on and off trains praying for signal failure because, if there is one thing that British trains are unreliable in, it is their unreliability. The best exposures are on the south side of the town at Old Maid Rock, which lies between the southern end of the beach and Coryton's Cove.

The verdancy of the market gardening areas around the coast is put into sharp relief by the intrusion of the granites that form the uplands of Devon and Cornwall. They were also a result of the Variscan Orogeny, responsible for the formation of the Ural Mountains that mark the border between Europe and Asia in Russia and the Pyrenees that straddle the French–Spanish frontier.

Despite the apparent separateness of Dartmoor, Bodmin Moor, Hensbarrow Downs, Carnmenellis, the Land's End peninsula and the Isles of Scilly, the individual characters of the moors are an illusion. All of these outcrops are united below the ground as one large, contiguous block of granite, though there is evidence for it being injected as sheets rather than bobbing up whole like a magma balloon through the crust.

The granite itself has an obvious effect on the landscape for it *is* the landscape, the sculptural tors dominating each and every view. In other parts of Britain, granite intrusions have developed a harder, sharper look about them – frost shattered, as they were, by repeated periods of glaciation – but on the south-western moors they have achieved a more rounded look than their northern counterparts. In part, their different appearance also stems from a certain sympathy with the landscape they are in. While the arêtes and peaks of glacially abused granite often appear to be there in spite of everything else working to remove them, the moorland rock is much more concordant with its surroundings, including horizontal joints that often appear as if they are bedding planes which, granite being an igneous rock, they cannot be. On hilltop tors the joints are usually horizontal; on tors that spring from the side of a hill, the joints often appear to follow the line of the hill, so that not only do they appear to mimic the structure of sedimentary bedding planes, but those planes bend and tilt as if part of a mock-anticline.

All is not what it seems, however. The heavily eroded and softer forms, along with those faux bedding planes, were hard-coded into their shapes at the moment of their formation. When the solid – but still hot – granite started to cool, it contracted, causing vertical joints to appear as a response to lateral forces in the crust. Mineralised water, heated by the still-hot rock, flowed through these fissures and formed minerals like quartz and tourmaline as a result. The erosion of the sandstone and slate that covered the granite – which in Dartmoor's case was between 6,000 and 10,000 feet, or between 2 and 3 kilometres thick – released pressure on the batholith which then expanded upwards, creating the lines of weakness that form the horizontal and tilting joints.

The lines of weakness were then exploited by various processes. First came the kaolinisation,[3] where some of the granite's softer minerals were broken down by hot water that continued to circulate in the rock for some time, because granite is naturally radioactive and generates its own heat. A period of deep weathering followed a long time after the kaolinisation; when Britain had an almost tropical climate the granite was attacked by acidic water caused by rotting plants. This water made its way along the fissures and, like the process of kaolinisation, favoured attack of the

---

3. Granite is mainly made of three minerals: quartz, which is very hard, biotite and feldspar. Feldspar makes up to 40 per cent of the granite and is decomposed by the action of hot water vapours from the depths to form a white clay known as kaolin, an important economic mineral as we shall see later.

softer feldspar minerals. The final blow, as with almost every rock in Britain, was the ice. Although not affected by glaciers per se, the south-west was periglacial and so the mechanical effect of freeze-thaw got to work on the tors. Summer thawing of the top layers of soil over the still frozen lower layers would then result in soil flow – or solifluxion – whereby loose material would slide en masse downhill. The boulder fields that lie under the tors were formed in this way; the geological term for it is the rather onomatopoeic-sounding clitter.

The joints of the tors not only mimic non-igneous works of nature with their pseudo bedding planes, but in times past have also been credited as the work of man – especially where massive boulders teeter on top of one another in such a fine balance that it seemed inconceivable that it was the product of dumb force. The Cheesewring on the south-eastern flank of Bodmin Moor is one such sculptural masterpiece – one which always reminds me of Zen stones, the pile of precariously balanced pebbles often used as a visual metaphor for meditation, an illustration of oriental mystic contemplation. The Cheesewring looks essentially the same, except larger, as if assembled by a team of preternaturally large Buddhist life coaches. Meanwhile, the nearby stones that were erected by Neolithic and Bronze Age people were turned by the early Church into signs of God's retribution for various acts of pagan activity – usually for dancing or playing music on the Sabbath. There are a number of stone circles, and one rare stone row, in Cornwall called Nine Maidens which, if the early Christian missionaries were to be believed, were where frolicking heathens were turned to stone by a not-so-much vengeful as spiteful God.

Though now largely uninhabited, save for a few small

towns and villages and an ugly, utilitarian prison at Princetown, Dartmoor was once at the cutting edge of Britain's industrial economy. Granite, once used for stone circles, stone rows and cromlechs, then hut circles, farms and cottages, is also a fine monumental stone and was used in countless high status city buildings and works of extravagant memorial like Nelson's Column, as well as more – if you'll forgive the pun – pedestrian use as a paving and kerb stone.

The manner of granite formation also led to one of the most distinctive kinds of human architecture to be found anywhere in Britain: the mine engine house. On the western end of Cornwall, where the granite of Penwith defines and controls the shape of the entire headland, the tin and copper mine houses with their solitary tall chimneys are such an iconic piece of Cornish identity that they have been appropriated by almost every enterprise that wishes to stamp their product with the county's version of the French *appellation d'origine contrôlée*.

Even if they were built out of a completely different rock, the tin and copper mines of Cornwall and West Devon would still be a paean to the formation of granite. Veins of minerals radiate out into the bedrock from the intrusions, effectively the fossils of superheated injections of water solutions rich in metals and rare elements. The same economic imperatives that have led to Chinese slate being nailed to the roofs of Cornish houses, despite a local abundance, crippled the Cornish tin- and copper-mining industries even earlier than the slate quarries. Barely limping in to the twentieth century, all the mines have now closed down, endowing Cornwall with the dubious honour of becoming the world's first post-industrial society. But the architecture required to support them remains, overrun by gorse and brambles in places, lovingly restored in others, perched on wild clifftops all around the north and west coasts of Cornwall.

The engine houses are quiet and still now, like monuments, but were at one time alive with the noise of Cornish beam engines pumping water out of the mines. Now that the reasons for their existence have all been removed, the engine houses – as much a part of the machine as the engine inside – might look like magnificent romantic follies, were it not for the fact that they serve instead as memorials; the surface indicators of subterranean hardship, of graft and gross exploitation. The Cornish mining industry, the profits of which were probably all well spent in cities miles away from the county, has, however, left a few artefacts of note.

The development of high-pressure steam engines – a hotly contested issue – owes a lot to a Cornish mine engineer, Richard Trevithick. It was he who first demonstrated the principle of a steam locomotive on Christmas Eve 1801 in Camborne, Cornwall. His invention was one of many breakthroughs driven by the needs of mining for more efficient pump engines and came a full twenty-eight years before Stephenson's *Rocket*. Yet he is only really honoured in his native Cornwall and in Cornish ex-pat communities around the world. There are many other fine inventions inspired by and used in the metalliferous veins of the Permian that gave rise to the tin- and copper-mining industries of Cornwall, but one of them is a particular favourite of mine: the rather low-tech, but delicious Cornish pasty. The pasty was designed for taking down the mine and, it is claimed, often contained breakfast at one end and lunch at the other. The crimp that runs down the spine of the pasty, by the way, was designed for holding while you ate, thereby keeping the pasty clean as well as keeping traces of arsenic out of your mouth. Mining tradition insists that the crimps were thrown away for the 'knockers', the capricious spirits who would knock on the walls before the mine collapsed, as they often did.

The formation and decomposition of granite also had another mineral by-product, kaolinite, the more familiar name of which is china clay. China clay is everywhere in the modern world, not just in china clay or porcelain. It is used as the coating in coated paper, in toothpaste and cosmetics. The white pearl lining of electric light bulbs contains china clay, it is used to reinforce rubber, as a food additive and has many other industrial applications. As porcelain, it doesn't just make your best tableware, but is also useful as an electrical insulator, for building materials, false teeth and objets d'art. When William Cookworthy, the Quaker apothecary and potter, began mining it at Tregonning Hill between Helston and Penzance in the eighteenth century he changed the landscape as well as the fortunes of the county.

After Cookworthy had made his discovery, it wasn't long before a much larger and potentially more lucrative deposit of kaolinite was found in the Hensbarrow Downs, north of St Austell. The china clay of St Austell is the largest outcrop of kaolinite in the world, is particularly suited to use in paper coating and has accounted for over 120 million tonnes since the first pits were opened. Furthermore, it is estimated that there are still a hundred years of deposits left. After nearly 250 years of extraction from Hensbarrow Downs, the china clay industry kickstarted by Cookworthy has turned them into one of the most distinctive landscapes in Britain, if not the world. To illustrate why, I'll give you their local name: 'The Cornish Alps'.

The locals don't call them that for the purposes of irony, but the name has been coined because of their appearance rather than their height.[4] In order to get the kaolin out of the quarries, it is washed by very high pressure hoses, but for every tonne of china clay produced, there are nine tonnes of waste. The alpine peaks are the eerily beautiful, roughly conical slag heaps or 'sky tips' of white spoil left over from quarrying activities. On a clear day you can see the sky tips from miles around but once you get close to them, you realise that they really ought to call the area the 'Cornish Moon', as it is one of the most surreal landscapes anywhere in the world – so surreal that it is a surprise it hasn't been more frequently used over the years as a TV and film location. Among the productions looking for scenery on the extra-planetary side of things have been the BBC's *Doctor Who – Colony in Space* in 1971 and the TV series of *The Hitchhiker's Guide to the Galaxy* in 1981.[5]

When a pit is abandoned or a new tip is started and a tip is left in peace, it quickly starts to green as ecological succession takes over with grasses and heathers, followed by gorse, but they still stand out on the skyline in all their conical glory. The white bases of recently flooded pits lend the water an odd blue-green Caribbean appearance in the

---

4. The highest peak of the Hensbarrow Downs, Hensbarrow Beacon, only rises to 1,024 feet (312 metres) even though the adjacent tip rises to 1,164 feet (355 metres) above sea level. Which only goes to show that, as a nation, the British are incredibly rigid and inflexible about what constitutes 'a hill'.

5. On film, the china clay pits also popped up in Richard Lester's 1969 post-holocaust satire *The Bed Sitting Room*.

bright sunshine – a hint at a lush tropical beach, but not the kind of cerulean sea you would fancy a dip in. These too develop ecologically over the years and become thriving habitats in a relatively short period of time.

Had they been left alone, unquarried, they may have looked today like the top of the West Penwith outcrop of granite – heathland and moor punctuated by carns, quoits,[6] chambered tombs and Iron Age hill forts. Of all the West Country granite bosses, this is the only one in direct contact with the sea. The very shape of West Penwith is the shape of the outcrop, surrounded on three sides by the Atlantic and joined at the fourth by a low and narrow isthmus to the rest of Cornwall.[7]

The overall landscape of West Penwith – moorland, isolated farmsteads and hamlets and a few large villages – does seem to be lodged right on the cusp of an industrial age that arrived but didn't hang around for long enough to overcome its geographical isolation. This is how most of Britain must have been before the development of canals, railways and roads centralised the infrastructure and set the modern pattern of settlement once and for all. In other parts of Cornwall, notably around the china clay area of the Hensbarrow Downs, large vibrant villages remain, each big enough to sustain local shops, like butchers and greengrocers that become exceedingly rare in small towns once

6. A quoit is a megalithic tomb with a large flat slab of stone resting on a number of upright stones. In other cultures, it is known as a dolmen or cromlech. The quoits, which date from the early Neolithic, were formerly covered in earth to make barrows.

7. Interestingly, even after living in the West of Cornwall for the best part of twenty years, I never discovered exactly where 'East Penwith' was and now believe that Penwith equals West Penwith and vice versa.

they find themselves colonised by the industrial-scale supermarket chains. In West Penwith, the idea of the community still thrives despite the economic conditions and the small town of St Just in Penwith, a little to the north of Land's End, remains isolated enough to sustain itself, living on the proceeds of marginal farming and a desperately short summer tourist season.

The tourists come to see Land's End, of course, the beautiful mile-long sandy beach of Whitesand Bay at Sennen and numerous coves and porths along the coast. They also come to see the remains of Bronze Age civilisation, the exceptional landscape art of local painters, either born here or drawn by a numinous light afforded by the peninsular nature of Penwith.

West Penwith is a truly unique landscape. As well as the iconic engine houses left by an industry that seemed to begin its decline at the time that the rest of the Industrial Revolution was getting underway, that part of the interior of the peninsula that has been reclaimed from the heather and moorland is enclosed in tiny and irregular fields separated by Cornish hedges. 'Hedges' in Cornwall may look soft and verdant but, as anyone whose car has bounced off a Cornish hedge with even a glancing blow will tell you, they are made of granite. Many of these walls are unspeakably old. In his 1986 book, *The History of the Countryside*, land-use expert and landscape historian Oliver Rackham wrote of them: 'In the Land's End peninsula there is one of the most impressively ancient farmland landscapes in Europe. The farmland is of tiny irregular pastures separated by great banks, each formed of a row of "grounders" – huge granite boulders – topped off with lesser boulders and earth.'

Rackham then goes on to make his most extraordinary claim, one that puts the landscape of West Penwith in a league of its own: 'The banks, from their construction, are contemporary with the fields. They can be roughly dated by the Bronze Age objects which were buried in the banks. These banks, indeed, are among the world's oldest human artefacts still in use.' Though a few archaeologists accuse Rackham of hyperbole, there is some local consensus that, especially in the case of the northernmost area of Penwith around Boswednack and Zennor, he is right. Defra and the Countryside Agency sanctioned the view – along with the County Council – and all went about encouraging farmers to keep the walls as they are, while most farmers were happy to oblige as the walls protect their livestock in this windiest of landscapes against the Atlantic gales.

The bronze that powered the Bronze Age, whose occupants built the field walls, is an alloy of the same tin and copper streamed and mined for thousands of years in the area. In turn, the zenith of the mining industry created the iconic engine house silhouettes on the moors and around the coast. It is perhaps fitting that all the features of this landscape are so interconnected, all an expression of the forces that formed it.

CHAPTER NINE

Millstones, Coal and Lime – the Peaks of Industry

Carboniferous period – from 299 million to 359 million years ago

Anyone with a GCSE or O Level in Geography will know something about the Carboniferous period, even if it is only a poorly remembered field trip spent gazing at cliffs in the Peak District in clothes that are slightly damp.[1] Springing to mind, possibly due to a school trip to Malham Cove as a disgruntled teenager in a rustling cagoule, is the karst formation of landscape, particularly that of limestone pavements.[2] For this is a landscape like almost no other

1. That is, still slightly damp, despite the maddening rustle and high-visibility optical pollution afforded by the average field trip cagoule, the overall effect of which is like standing next to someone packing away an infinitely complex tent while shouting obscenities into your eye.

2. A limestone pavement, if you don't remember your school geography, is formed when a flat bed of limestone is attacked by slightly acidic rain along joints and cracks in the rock. (Even before the industrial era, all rain was slightly acidic, by dissolving carbon dioxide from the atmosphere to form a weak form of carbonic acid.) The acid widens the cracks into fissures called grykes, leaving slabs known as 'clints' standing proud. (cont. p. 149)

Outcrops of Carboniferous rocks in Britain

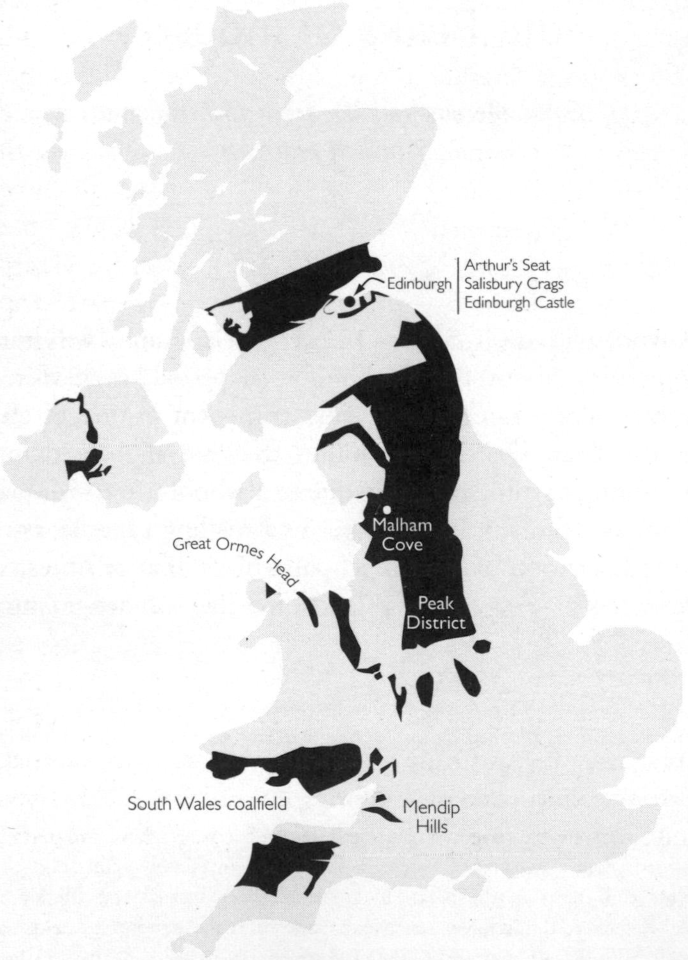

and even the underlying geology has its own underworld of subterranean passageways and caves.

Over the preceding chapters, it has become clear that topography does not always follow the geology in the most obvious ways and that the connection between the landscape and the underlying structure is often more abstract and indirect. Limestone bucks the trend, however. The nature of the Carboniferous limestone is perhaps best hinted at by its alternative, slightly anachronistic name of 'mountain limestone'. If the Pennines are known as the backbone of England, likewise then the mountain limestone that predominates on England's peaks is the spine of the later geology – the Jurassic and Cretaceous escarpments that run in a garish south-west–north-east stripe across the geological map of our islands from Dorset to Yorkshire. To the west of these outcrops, it marks a transition between the rugged moorland of the north and west and the softer lines of the south and east.

Outcrops of Carboniferous limestone form deep gorges and rocky upland. In Britain it tends to occur along a broad line drawn from the Mendip Hills of north Somerset, where the mapmaker William Smith acted as a mineral surveyor; around the Bristol Avon gorge; encircling the South Wales coalfield; along the Wye Valley; in the Peak District; and in a few parts of Anglesey and North Wales – including Great Ormes Head, Llandudno and the magnificent escarpment of Eglwyseg, near Llangollen. Inevitably, however, one of the best places to see the mountain

---

The grykes are oftenhome to woodland species of plants, like honeysuckle and dogs mercury – all that remains of the original woodland that would have covered these areas before clearance for grazing.

limestone is in the mountains and extensive outcrops of it can be seen not only in the Peak District, but further north in the Pennines, particularly in the Yorkshire Dales.

Unless it is actually raining – and that is a far from rare occurrence in British field geology – there is a general absence of surface water on the limestone pavement overlooking Malham Cove, which is odd because the formation of the 260 foot (80 metres) inland cliff you are standing on is entirely the product of water. Some of that water, admittedly, was in its solid and most destructive form, glacial ice, doing all the things that glaciers do: gouging, scraping, freeze-thawing, but most of it just fell out of the sky in the normal way. So, where exactly is all this water? Looking down to the cove below provides the answer, where Malham Beck emerges from the foot of the cliff. It is all underground.

The cliff of the Cove was, in glacial times, the site of a waterfall that drained Malham Tarn, a few miles to the north, which at 1,240 feet (377 metres) above sea level, is England's highest lake. The Tarn lies on an outcrop of impermeable Silurian slate – only a few times larger than the lake itself – and was dammed by a ridge of moraine deposited at the edge of a glacier during the last period of glaciation. While the ground was frozen and rendered impermeable, the stream flowed towards and over the cliffs of the Cove, but once the ground thawed and the limestone became permeable again, it started to slowly dissolve the limestone to create the sinkhole now known, with typical Yorkshire directness, as 'Water Sinks', leaving a dry valley (called 'Dry Valley') behind. Today, a steady stream of walkers are the only flow – the Pennine Way runs along the Dry Valley and thousands of feet grind against the limestone in place of the swirling waters. Only in extreme flood conditions does the stream flow down the dry valley to cascade over the cliff at Malham Cove and join the beck at its foot – the last known instance when the water table was high enough was in the early nineteenth century.

Given the prehistoric drainage pattern, it would be tempting to think that Malham Beck itself, the stream that emerges under the cliff of the Cove, is fed by the stream that disappears at Water Sinks, but this isn't the case. The system of caves is much more complex than that as experiments with dyes have conclusively shown. Somewhere in the extensive caves below the moor, Malham Tarn's water crosses the path of the waters that feed Malham Beck without any confluence occurring and then emerges some way south and downstream of the Cove.

Goredale Beck joins the Malham Beck further downstream still. A visit to Malham Cove usually takes in the neighbouring feature of Gordale Scar just over a mile (1.75 kilometres) to the east, from which the Beck emerges in two waterfalls. The Scar – at over 330 feet (100 metres) high – may have once been a cave that suffered a collapse of its roof and, in places, the cliff does seem to overhang, adding to the feeling of oppressive power. James Ward famously made it the subject of a brooding masterpiece of English Romantic painting in 1815; a masterpiece which makes it clear just how insignificant human affairs are against the backdrop of almighty nature. The effect of the painting, although magnified by romance, is truly stunning and comes the nearest of many artistic attempts over the years to convey the sheer brute force of such a landscape. J.M.W. Turner sketched it in oil and pencil the following year and, never far behind in the race to immortalise the sublime grandeur of nature, William Wordsworth wrote of it in a sonnet in 1818.

> Then, pensive Votary! let thy feet repair
> To Gordale-chasm, terrific as the lair
> Where the young lions couch; thou may'st perceive
> The local Deity, with oozy hair
> And mineral crown, beside his jagged urn [. . .]

However, the mountain limestone of Gordale and Malham was not the only activity of the Carboniferous. The period was one of enormous change: oxygen concentrations more than doubled from the previous period – the Devonian – to 35 per cent of the atmosphere at their peak in the Carboniferous (compared to today's concentrations of approximately 21 per cent). In the meantime, carbon dioxide levels had fallen by two-thirds. The Carboniferous saw Britain switch between a shallow tropical sea and a sweltering swamp.

In the oxygen rich atmosphere, insects were set free of physical limitations to their body sizes – one dragonfly-like predator, *Meganeura*, had a wingspan of 2.5 feet (75 centimetres) and was the world's largest-ever flying insect. It does not take a huge leap of the imagination to visualise *Meganeura* flitting around a warm swamp seizing an insect or small amphibian and, indeed, fossils of it appear in the rock strata laid down at the time – the coal measures.

Carboniferous means 'coal-bearing', as named by the geologist Reverend William Conybeare in his 1822 book, co-authored with William Phillips, *Outlines of the Geology of England and Wales.* The name of the period which bridged the gap between the New Red and Old Red Sandstones has stuck, despite Conybeare's initial use of it as an epithet, noting as he did the presence of the mountain limestone and other rocks of what he preferred to call the 'Medial Order'.[3]

During the early Carboniferous, the equator straddled the north of Britain, but much of the south of the country lay under water. A lot of this water was shallow, an epicontinental sea rich in corals that flooded the continental shelf and responsible for the deposition of the mountain limestone of the Pennines, South Wales and the Mendip Hills. Between the shelf seas of the Pennines and those of Wales and Somerset, a broad finger of land extended from the east over mid Wales and into the Irish Sea, now known as the Wales-London-Brabant Massif, but which used to go by the much friendlier name of St George's Land. To the north of the Pennines – a line roughly along the Scottish border – there was more dry land on the edge of the Old Red Sandstone continent, interrupted only by areas of lakes and river deltas around the Midland Valley of Scotland.

In the later Carboniferous – with the climate becoming steadily more humid – the shallow sea areas became, by then, vast areas of swamps and rapid plant growth, the results of which we burn in our power stations now. Like oil, coal is a geological accident. While the atmosphere was rich in oxygen, the production of coal demands anaerobic conditions – oxygen allows fungi and bacteria to break down organic material which forms soils. The soaked branches and trunks of the Carboniferous swamps were

3. In North America, the Carboniferous is split into two epochs – the older, longer Mississippian, dominated by limestone deposits, and the shorter Pennsylvanian, which, in North America, contains the coal measures.

buried in mud very quickly as whole forests were inundated by the sea and slipped under the weight of sediments into basins. Gradually a new forest grew on the new soil and the whole process is repeated. As layers of sediment build up, pressure and heat slowly metamorphose the organic matter into coal, forming thin seams within the measures which also contain sandstones or shales formed from the sediments that originally buried the forests. This wholesale burial of plant matter acted as a carbon sink until man started burning it a few thousand years ago.

The swamps were founded on broad river deltas that had formed as the outflow from sediment-rich rivers washing down from the nearby highlands. The deltas eventually became a special type of coarse sandstone, prevalent all over the north of England, called millstone grit. Millstone grit forms as bleak a landscape as it sounds. So much so, that it is often mistaken for granite – and was probably formed from it, sharing, as it does the quartz and feldspar which make up the larger part of the igneous rock. Whatever the reason for the confusion, no lesser a figure than Charlotte Brontë made just such an error in *Jane Eyre*: 'I struck straight into the heath; I held on to a hollow I saw deeply furrowing the brown moorside; I waded, knee-deep in its dark growth; I turned with its turnings, and finding a moss-blackened granite crag in a hidden angle, I sat down under it.'

Putting aside the feeling that Brontë's writing would have suffered as a result of deeper geological understanding it is, nevertheless, an easy mistake to make. Like granite, millstone grit forms tors and the rather uncompromising and bleak moorland of the kind beloved by an author of haunting romance.

You can still find quarried grit under Stanage Edge in the Peak District National Park, carved into millstones and ready to grind barley, but which were never carried away. It was also used as a building material for the mills themselves. For all its butch-sounding qualities and its provenance from the granite, millstone grit is rather susceptible to weathering. Ten miles north-west of Harrogate in the Yorkshire Dales National Park, Brimham Rocks[4] show the very zenith of rock weathering, where the forms on display approach abstract sculpture. Over an area of around 50 acres (20 hectares) of Brimham Moor, the curious forms are everywhere. Softer layers of the grit have been scooped out in places and, in the case of the Idol Rock, 200 tons of rock sit on an almost comically tiny pedestal.

These three rocks – millstone grit, coal and mountain limestone – almost always found together in Britain, sum up the bulk of the Carboniferous. In some parts of the Pennines there is a kind of cycling of repeated successions of rock, for instance from limestone, to shale, to sandstone, to limestone again and so on. Sometimes a thin seam of coal gets caught up in one of these repeated successions – called cyclothems – but the general theme of constant reiteration is the same, suggesting a pattern of repeated marine transgressions and regressions over the course of the Carboniferous.

The Carboniferous, then, was a period that had such a split personality. On the one hand there is the subterranean world of caves, stalagmites and stalactites, of swallow holes

---

4. Can be very busy on holiday weekends with queues for parking, but ultimately worth the slightly ludicrous parking charge.

and streams lost to the percolation of water, while on the other there is the stygian gloom of coal mines with their industrial landscapes of slag heaps, colliery winding gear and similar utilitarian structures – the Pennsylvanian bits of our most recent heritage we have the most problems reconciling. Between them was the millstone grit, formed in massive river deltas which would later be the dank forest floor of a tropical lagoon, inundated by the sea time and time again. In North America, as has already been noted, the character of the Carboniferous is so split that it forms two entirely different sub-periods: the earlier, longer Mississippian of limestone and the later, shorter Pennsylvanian of the coal measures.

There are two final landscapes of the Carboniferous, one of which inspires fascination and awe among primary schoolchildren who are just starting to learn about the surface of the earth, the other of which is less well known. To see both this iconic landform and its lesser known but related feature, we must travel north to Edinburgh. We must climb a volcano.

I elected to climb Arthur's Seat one early Sunday morning in March after a bout of traditional Scottish drinking the night before. On reflection, I think I might have still been a little drunk because, although it is a moderately strenuous and mercifully short hike by climbing standards, an overweight asthmatic should probably prepare himself a little better for the 820 feet (251 metre) ascent. Part of my overconfidence undoubtedly lay in the simple fact that Arthur's Seat is essentially a city centre mountain and I became fixated in the belief that if there wasn't actually a bar at the top, there would at least be an ice-cream van or refreshments kiosk somewhere en route.

I arrived at the summit in what I am going to euphemistically call a crumpled state, but the view was more than worth it and the chilled wind sorted me out.

Rather than a mountain placed within a city centre, Arthur's Seat stands above Edinburgh in a manner that suggests that it is the other way around. Removed from the overwhelming influence of urban convenience and human-scale affairs, the city seems small by comparison to the geology, making it feel more like a city on the slopes of a mountain.

That mountain is what remains of a volcanic vent that burst through the calm tropical shoreline of Carboniferous Edinburgh. It started at what is now the Castle, standing on cliffs of the indomitable basalt rock it left behind. In short order, a new vent was forced through the lagoons in violent eruptions a little to the east. Tons of ash were thrown into the air and lava ran over the plain, while a cone of enormous proportions was augmented with yet more lava and ash with each new eruption. Each of Edinburgh's seven hills – which the city, like Rome, is famously built upon – were built themselves either by volcanic or other igneous processes. Not only Arthur's Seat, but Castle, Calton, Corstorphine and Craiglockhart, Blackford and the Braid Hills are all formed of resistant rock which held its ground in the face of massive glacial erosion during the Devensian glaciation.

The products of Edinburgh's geology of fire and ice can be seen from any of the commanding heights of the volcanic and igneous outcrops that make up the city. They include the classic 'crag and tail' arrangement of Edinburgh Castle itself. Selective glacial erosion of the softer sedimentary rocks in the area left the volcanic and other igneous rocks standing as hills in the landscape. The ice moved roughly west to east through Lothian and all the hills left standing have a 'lee' side tail to them. When glaciers moved over the crag of Edinburgh Castle, it protected the tail of softer sedimentary rocks that form the ridge that the Royal Mile now lies upon. The medieval Old Town of Edinburgh was confined to the easily defended crag and tail of the Castle Rock and the limited space had an early effect on development, leading to ten-storey 'high-rise' residential buildings.

In the late eighteenth century, expansion of the city beyond the crowded Old Town became inevitable and the New Town was founded to the north of the castle, again on a ridge, this time one that runs along a glacially produced, whale-shaped hill known as a drumlin. The ridge of the drumlin and the centre of the planned development – amongst the finest Georgian architectural treasures of Britain – runs along the magnificent George Street, between Charlotte Square and St Andrews Square. Meanwhile, Princes Street Gardens, which occupy the space between the Old and New Town, were gouged out by the glaciers diverted by the crag and tail of the Castle Rock, the hollow becoming Nor Loch, which was subsequently drained.

It seems fitting that, while the New Town was being planned and built, the Scottish Enlightenment was constructing an entirely New World and Edinburgh was at its centre. Thanks to the first public education system in Europe since the classical empires, Scotland had the highest rate of literacy in the continent and following the Act of Union in 1707, it was suddenly free to export its ideas to the world. Voltaire once famously said that 'We look to Scotland for all our ideas of civilisation.'

Among the famous figures of the Scottish Enlightenment – Adam Smith, David Hume, James Watt and Robert Adam – there was one who had built his home near Salisbury Crags, under Arthur's Seat: James Hutton, the man who has since been described as the father of modern geology. The approbation is fitting; it was Hutton who first articulated the idea of deep time; it was he who first described the unconformity, a staple of geological thought; and it was Hutton who first put forward the idea that the interior of the earth was molten and that rocks did not just precipitate out of a Noachian Flood, but also could be shown to have crystallised out of magma. One of the sites where he could show that this had happened, indeed, was Salisbury Crags, a part of which has now been named as 'Hutton's Section'.

Hutton's Section shows how magma was forced between bedding planes of sandstone, forming an igneous intrusion that geologists call a 'sill', tearing and rotating the bedrock as it squeezed its way through. By the very nature of Salisbury Crags, Hutton was able to demonstrate that the rocks of the earth had not formed all at once by one process, but were part of a much more complex and longer train of events. The intruding rock of Salisbury Crags is dolerite, which is chemically equivalent to the basalt of the volcanic vents of Arthur's Seat and Castle Rock, but which, because it never made it to the surface, was therefore much slower to cool – allowing for larger crystals to grow. Next to Arthur's Seat and the Castle Rock, Salisbury Crags are the most obvious landscape feature of the whole of Edinburgh. They are a magnificent and distinctive part of the city's skyline, but there is an even more pronounced dolerite sill, formed during the Carboniferous period and in exactly the same way as Salisbury Crags. This sill is much further south, in England, to be precise, though it still has a Scottish connection of sorts. It's called the Great Whin Sill.

In terms of effect on the landscape, there are not many geological features in these islands or elsewhere which have had such a marked influence on the landscape as the Great Whin Sill. The sill, which was injected into the surrounding bedrocks of Carboniferous sandstone, limestone and shale towards the end of the period around 295 million years ago, outcrops in a number of places in Northumberland and Teesside and is responsible for some of Britain's most stunning scenery.

The sill underlies much of north-east England and is responsible for, amongst others the High Force and Low Force waterfalls, the Farne Islands and, for several miles around Housesteads Fort, it forms a formidable north-facing cliff that augments the defences of Hadrian's Wall. The Great Whin Sill is also the original sill – local quarrymen used to call any horizontal bed of rock a sill and, when geologists worked out how the Whin Sill was formed, they appropriated the name to apply to all similar igneous intrusions.

The Carboniferous was a unique period in the formation of our landscape. Leaving aside the mountain limestone and the millstone grit and even the volcanic and igneous features, no other period accounts for the production of so much coal. That industry and its subsequent decline has had its effect on landscape as surely as it has affected the regional economies that depended upon it, though the industrial ruins may never mellow enough to be regarded as picturesque in the same way as the mine engine houses of Cornwall and Devon.

CHAPTER TEN

Iapetus, Caledonia and the Highlands Controversy

The Caledonian Orogeny – from 359 million to 430 million years ago

When I was told the story of the old lady from Caithness, who shooed off rescue helicopters with a broom in the blizzards of 1947 because she believed her rescuers were from another planet, I was in a postbus in the North West Highlands, gawping at a landscape which seemed distinctly alien. While I concede that I was on planet Earth, that part of Scotland certainly didn't feel like Britain in the slightest. A feeling that is not so far-fetched, because Scotland is not, in a geological sense, part of Europe at all – indeed, it was once part of the same land mass as Greenland and North America.

You get a subjective sense of this 'otherness' in the North West Highlands where towering Torridonian mountains rise from the Lewisian gneiss. It is almost as if they remember their American provenance and bear superficial similarities in their comparative scale to the mesas and buttes that tower above the Colorado Plateau – the John Wayne location of choice for all 'cowboy and indian'

The Moine Thrust and the North West Highlands

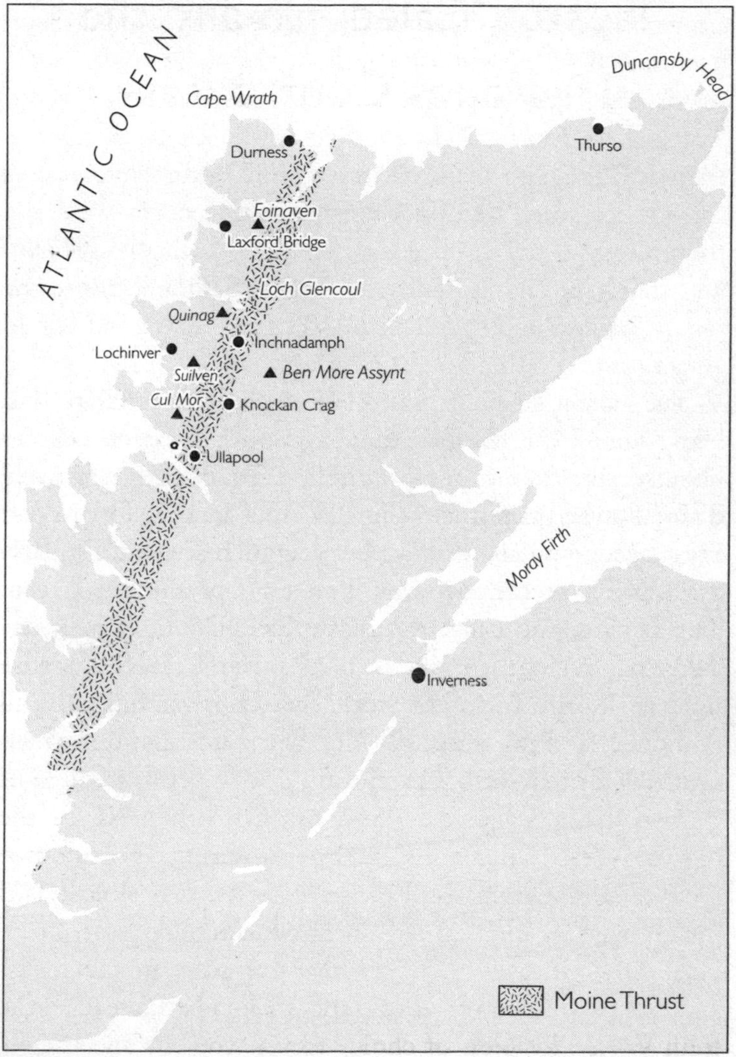

films before *A Fistful of Dollars* marked the start of the spaghetti western period. There are also some superficial similarities between the series of rocks that make up the Torridonian landscape and their counterparts in the Grand Canyon in Arizona. Despite the fact that they would have been laid down at opposite ends of the same continent, it seems that they experienced many of the same events. Those Torridonian peaks are even made out of continental America; they are mainly formed from sediments washed from a landscape that once existed to the north-west, but there are also traces of gneiss from the Canadian Shield[1] and Labrador 500 million years before Scotland joined the rest of Britain and two land masses from either side of an ancient ocean – the Iapetus – became one.

I've taken a liberty with our geological timeline and leapt back a bit further than I should for this chapter because the events of 430 million years ago effectively dictated everything that followed. The Iapetus Ocean had kept a geographical distance between Canada and England but also, it seems, between Preston and Gretna Green. The Iapetus was a proto-Atlantic Ocean[2] and once separated the ancient continent of Laurentia (analogous to modern North America) from Baltica (Scandinavia) and Avalonia (a thin strip of Northern Europe on which southern Britain lies). The closure of this ocean resulted in

1. The Canadian Shield is the ancient nucleus of North America, contains the oldest rocks in the world, and is arranged in an almost circular U-shape around the Hudson Bay area of Canada.

2. Hence the name. In Greek mythology, Iapetus was the father of Atlas, after whom the Atlantic was named. Some scientists still refer to the Iapetus Ocean simply as the proto-Atlantic.

a monumental collision of land masses, land masses on which Scotland and southern Britain were once as much as 4,000 miles apart. The collision formed a super continent variously named as Laurussia or Euramerica which, eventually, became part of a grand accumulation of all the world's continents, Pangaea.

The final and most fitting glory of this blunt and brutal movement of crust around the globe is that the Iapetus Suture – the line that marks where, as far as can be determined, Laurentia and Avalonia join – runs almost exactly along the English–Scottish border. When Scots speak of the uniqueness of their homeland, perhaps this is the ultimate proof. Perhaps we should go easier on American tourists who express their heritage in an unwise predilection for tartan and tales of sheep rustling over a wee glass of malt. For them, gcology as well as genealogy might show that Scotland is closer to home than perhaps we thought.

With three continents colliding at differing speeds, a lot of the actual buckling and movement of rock is difficult to decipher, particularly in the Highlands. Much of the ancient landscape is hidden under a block of rocks, some of them of indeterminate age, that were continually thrust over it during the fusion. This feature, known as the rather butch-sounding Moine Thrust, is formed from schists[3] which have been heaved 40 miles (64 kilometres) over the

3. Schists are a metamorpic rock distinctive in their appearance because of small wrinkles and corrugations on their surfaces. The Moine schists may have been metamorphosed three times in the 500 million or so years between their initial formation and their wholesale movement as a result of the Moine Thrust.

landscape from the east, and then extensively deformed, folded, overturned and sheared off repeatedly and now stacked up over one another as a landscape of imbricated blocks, like toppled dominoes or the slates on a roof. Even detailed geology textbooks are a little vague about the exact sequence of events, but the broad sweep of what occurred is now agreed, even though it took over fifty years to get to that point. The eminent scientists of the nineteenth and early twentieth centuries eventually unravelled the geological Rubik's cube that is the North West Highlands and, in doing so, began to understand a landscape that was formed at the hands of almost incomprehensible might and force; a landscape which, at times, defied the rules that had already been laid down by a young science which sought to explain it all.

The thrust fault that was discovered is only the extreme result of folding, a process that had already been observed and understood by geologists at countless locations during the nineteenth century. When the pressures that produce anticlines and synclines are not roughly equal on both sides the fold becomes asymmetrical. More radically unbalanced forces produce overfolds, or nappes, some of which may be so lopsided that the axis of the fold becomes virtually horizontal.[4] Eventually, further compression may overcome the structural integrity of the rock itself and the upper limb of the fold will shear over the lower limb, creating a thrust fault.

Most of the guiding principles of geology were worked out in the eighteenth and nineteenth centuries and much

---

4. When the axis of a fold approaches the horizontal it becomes a 'recumbent' fold.

of that important work was done either in Scotland or elsewhere by Scottish scientists. The cutting edge, however, was in the North West Highlands, an area where the finest minds were drawn in to solve the most convoluted puzzles of all. Before the mechanisms that drove it were discovered, the peculiar succession of rocks that outcropped over hundreds of square miles of Scotland was a puzzle so intractable that even the scientists of the government-funded Geological Survey referred to it as the 'zone of complication'.

Much of nineteenth-century science was about collection and classification. Amateur natural historians of the day, for instance, would embark on long walks in the country punctuated by painting, shooting and egg collection, with periods of taxidermy and taxonomy in the study later to make sense of the day's spoils. Geology in its Victorian heyday was no different to other fledgling sciences. Beyond the collection of samples and drawing of rock formations in the field, much of the work lay in ascertaining the order of things and categorising and classifying layers of rock into discrete systems.[5] The language that accompanied the publication of their endeavours is quite revealing; once ordered and grouped, they 'erected' the system, as if it was a permanent edifice of knowledge, an unconquerable cliff face that would stand for ever, one immutable fact piled on top of another. The initial story of the North West Highlands, particularly that 'zone of complication', turned out not to be quite as edifying.

5. While geological time is divided into various units, the rocks themselves are divided into 'systems'. A system is analogous to a geological period: the period describes the time, the system describes the geology itself.

It all started with Sir Roderick Impey Murchison, a man who came to geology late after a career in the Army, yet who still had time to write over a hundred papers and books. Murchison was, without doubt, one of the pioneers of geology but was also a master of getting ahead of the pack, of setting himself up for greatness. In his excellent book *Scientist of Empire*, author Robert A. Stafford more or less encapsulates what we think of the man today when he describes the activities of Murchison's formative years in geology as: 'an impressive record of research and a relentless campaign of self-promotion'.

Between the 1820s and 1860s, Murchison made five visits to the North West Highlands, first with Adam Sedgwick – an early mentor of Charles Darwin – in 1827, then again in 1855 with James Nicol, Professor of Natural History at the University of Aberdeen. Nicol and Murchison disagreed with heart-felt passion about what they found there.

By this point in his career, Murchison had erected his very own system, the Silurian, which had, so far, successfully classified a series of rocks in South Wales and along the Welsh Border with a distinctive set of fossils all of their own – quite unlike anything that had been examined in detail in the British Isles up to that point. His Silurian series established, when he went to the North West Highlands Murchison noted a series of metamorphic rocks that lay over what he believed was a Lower Silurian sequence and recruited them into his Silurian system, effectively annexing the whole of the north of Scotland's rock formations into his geological expertise. Not for nothing was he known as the 'King of Siluria'.

Nicol begged to differ. In particular, he had great

difficulty with Murchison's metamorphic series, now known collectively as the Moine schists, lying slap-bang on top of unaltered limestone, among others. If Murchison was correct, what he described as micaceous and gneissose schists had been baked and pressurised in situ without any effect on the limestones mere inches away.

Nicol took issue with Murchison publicly in the scientific journals of the day, but he came up against the might of the establishment; as the son of a minister from Peeblesshire – a mere one-time assistant secretary and librarian for the Geological Society – he was no match for Sir Roderick Impey Murchison, who was by then the director general of the Geological Survey, vice president of the Geological Society, three years from becoming a Knight Commander of the Order of the Bath, at the top of his profession and gathering letters after his name at an astonishing rate.

Faced with such an imbalance of power, Nicol could easily have given in and acted with the deference and respect that was characteristic of Victorian society, but what Nicol lacked by accident of birth and status he made up for with a terrier-like grasp of his own convictions. Murchison, meanwhile, regarded Nicol as something of an annoyance and the two sparred in drawn-out debates summarised by the academic publications of the time. Both the Nicol and Murchison view of the North West Highlands are featured in the first part of the *Quarterly Journal* of the Geological Society of London in 1861. Nicol's geological memoir *On the Structure of the North-Western Highlands* was published alongside Murchison's *On the Altered Rocks of the Western Islands of Scotland and the North-Western and Central Highlands.* Both papers trade softly spoken lacerating scorn on one another's theories, but the authority of Murchison – as well as an easy style of writing, due in no small part to Murchison's co-author Archibald Geikie – won over the substance of Nicol and, as a result, the Murchison–Geikie interpretation of the North West Highlands became the official 'Survey Interpretation'.

Over twenty years later, the Professor of Geology at Mason's College, Charles Lapworth, had a hand in posthumously overturning Murchison's Silurian ambition. In 1883, four years after he had settled an earlier debate which had turned into a bitter dispute (over Murchison and Adam Sedgwick's Silurian and Cambrian strata in Wales) he went on, in large part, to settle what had become known as the Highlands Controversy in his paper *The Secret of the Highlands.* After an energetic survey – something which, it could be argued, the Survey itself had so far failed to undertake – Lapworth proposed that Murchison and Geikie's Silurian schists were nothing of the sort, but were instead much older rocks, counter-intuitively lying on top of younger ones. Lapworth then went on to describe what he termed as a 'low angle reverse fault' – meaning the rocks had been pushed over them – dipping to the south-east. Lapworth had described what later became known as the Moine Thrust.

There were other geologists in the field coming up with the same conclusions. Charles Callaway is often overlooked but it was he who discovered the Glencoul Thrust, a subsidiary movement to the Moine, where a sheet of rocks, including Lewisian gneiss, has been lifted up in the east and thrust westwards for 20 miles (32 kilometres) over younger strata which, in turn, lie in their original position on the gneiss of the foreland.[6] If you stand by the A894 at Unapool on the shores of Loch Glencoul, the giant slab of the thrust is all too apparent even without the benefit of knowing what to look for. It is even rather obvious on the Ordnance Survey map of the area where, around the peninsula of Aird da Loch, the structure is rendered as a wedge-shaped crag and a fearsome set of contours that make you want to throw your walking boots away.

Even though its outcomes were truly international, the roots of the Highlands Controversy were a rather British affair. With the benefit of hindsight, it's plain that Murchison and Geikie's ideas about the North West Highlands were not so much part of a scientific discourse or a living debate, but a doggedly defended position; an ossified, immutable fact that had woven itself into the fabric of the institution like the fossil of some ancient animal forever embedded in the bedrock.

In 1883, Lapworth's fine and unerring fieldwork and Callaway's observations eventually led to the Geological Survey itself, now presided over, ironically, by Geikie, to dispatch its best surveyors to the area to settle the matter once and for all. Geikie instructed the Survey's men, Ben Peach and John Horne, 'to divest themselves of any prepossession in favour of published views, and to map existing facts in entire disregard of theory', which looks generous for a man who had invested so much of his own reputation in the Survey interpretation.

Peach and Horne's investigations quickly confirmed Lapworth and Callaway's theories that older rock had been

---

6. Foreland is a geological term that identifies the unyielding block within a fault – the one that stays put.

thrust over younger at a very shallow angle – an angle so low that it resembled a largely undisturbed bedding plane to Geikie and Murchison. It was surely Murchison's instinct for augmenting his Silurian kingdom that was probably as much to blame as the deceptive appearance of the formation, because James Nicol, though his theory wasn't correct in every detail, had seen it quite clearly, more or less for what it was, in 1855.

The Survey moved swiftly to show interim findings in 1884 while Lapworth was ill, using the unconventional route of publication in *Nature* magazine. Lapworth once complained that he felt the Moine grating over his body as he lay in bed and it may be that the sheer vigour with which he attacked his Highlands work was to blame for the breakdown in his health.

In 1907, over twenty years after the interim findings, Peach and Horne published their seminal memoir *The Geological Structure of the North-West Highlands of Scotland*, a mercurial work of surveying and synthesis that did not just benefit those interested in one small, if complex, area of Scotland but was, rather, of global importance. Their discoveries and conclusions have since proven, along with the plate tectonic theory that provided the mechanism for the Moine Thrust, to be the engine of earth science for the entire planet. It was so revolutionary that the Alpine geologist Edward Suess remarked that the work of Peach and Horne had made the mountains transparent. A monument to mark the work of Peach and Horne stands in the tiny hamlet of Inchnadamph under Stronchrubie Crag, a feature of the Moine Thrust. It reads: 'To Ben N Peach and John Horne who played the foremost part in unravelling the geological structure of the North West Highlands

1883–1897. An international tribute. Erected 1980.' Lapworth, Callaway and Nicol are not even mentioned.

I visited Inchnadamph once as part of an exploratory tour of the area; it isn't that far, after all, from the Torridonian sandstone and Lewisian gneiss I went to see. In fact, the tiny hamlet finds itself sandwiched between all three and is an apposite place to consider the landscape for that reason alone, rather than the memorial to Peach and Horne, or even the Inchnadamph Hotel – famous in science the world over and a Mecca for geologists. Conditioned by the moss, rust and battleship palette of the Lewisian landscapes and the various slate greys of the mountains on the Moine, I entered into the valley at Inchnadamph and was reminded instantly, in the suddenly verdant landscape, of the colour green. After only a day wandering around the North West Highlands, it seemed more green than I remembered and the soft lines were greeted as a long-lost friend after the austerity of the surrounding scenery.

Inchnadamph lies on a long strip of limestone that stretches in a thin line from the north coast for almost 100 miles (160 kilometres) along the edge of the Moine Thrust. From Inchnadamph, by the side of Loch Assynt, you have views of both the Torridon peaks and the great thrust faults formed when Scotland and England were joined in the original act of union. Such was the upheaval of the collision, it would be easy to believe that all the mountains of the North West Highlands were involved, but the peaks of Slioch, Torridon and Suilven, among others, were beyond the north-western boundary of the orogeny and were left completely untouched by the tectonic forces that built the Caledonian Mountains.

However, to the east and south of the Moine Thrust, the picture is very different. It isn't only the North West Highlands that have confounded geologists for centuries. The rest of the country – the crumple zone of the Caledonian mountain-building event – is as intricate and difficult a geological structure as almost anywhere on earth. As if geologists just can't put down a good controversy, the origin of the rocks that were thrust over the vast area, the Moine schists, are still contested. There seems some agreement, although far from unanimous, that some of them are metamorphosed Torridonian sandstones. Others contend that the original rocks are pre-Torridonian, but post-Lewisian, filling a 2-billion-year hole in the geological record. Radiometric dating, which should have settled the argument, came up with no single date and showed that the Moine schists are – like the Lewisian gneisses – not a single rock at all, more a collection of different rocks with nothing in common but the circumstances of their metamorphism, their essential 'Moineyness'.

A geologist once guiltily confided to me that he thought that the Torridonian mountains of the North West Highlands – like Suilven, which we have encountered – were 'nicer' than the features formed later as part of the Caledonian Orogeny including, I assumed, the peaks of the Moine Thrust Belt itself. He had a point: the character of what are two entirely separate ranges in close proximity to one another is markedly different. The almost sculptural Torridonian inselbergs rising from the rolling boil of the Lewisian hummocks suggest their Laurentian origins, whereas there is something quintessentially Alpine about the peaks of the Moine Thrust. Nowhere can this be better seen than from the A837, at the western end of Loch Assynt on the road from Lochinver. To your right, the distinctly odd Suilven sticks up like a headstone in a forgotten graveyard, while straight ahead rises the massive wall of Glas Bheinn Assynt. Though nowhere near as tall as its neighbours to the south, Ben More Assynt and Conival, the impression of a broad peak straddling the horizon is much the same. Incidentally, there is a beautiful economy with which these three peaks are named. Glas Bheinn translates from the Gaelic as 'big bluish-grey hill', Ben More as 'big mountain' and Ben More's adjoining peak of Conival – in its original Gaelic form of *Conamheall* – as 'adjoining hill'.

By virtue of the thrust fault that had a hand in its formation, we can see that the most outstanding landscape feature of all the Moine rocks are the inland cliff faces and crags that run here and there along its length, thrust there by gigantic forces, but now still and facing the implacable terraces of Torridonian sandstone hills a few miles to the west. A few miles south of Inchnadamph near the hamlet of Elphin is Knockan Crag, the location of an unmanned 24/7 visitor centre in the shadow of the Moine Thrust. The centre is an integral part of the Scottish Natural Heritage Rock Route. As an educational attraction, it is superb; one of the reasons that it works so well is that it lies next to the very thing it seeks to explain. All that is needed to understand the enormity of what Knockan Crag represents – the tectonic bulldozing that occurs when two continents crash into one another – is a context board or two and a little sign to tell you that 'this is the place where one enormous block of rock was pushed over another'. If only the rest of Scotland were so simple.

In terms of the very ancient rocks of Scotland, the Moine schists stretch away to the south and east all the way to the Great Glen and beyond. From there on until a line that runs approximately from Stonehaven to Greenock, they are gradually superseded by an extensive group of rocks known as the Dalradian Super Group. The Dalradian, like the Moine, is another group of metamorphic rocks where, along with the schists, we find some better-known products of metamorphism, such as marble, which comes from limestone, and slate, which is the metamorphic descendant of shale. As the Iapetus Ocean closed the original rocks (protoliths) of both the Dalradian and the Moine were folded and then thrust over younger rocks.

The hundreds of square miles of contorted, metamorphosed rocks represented on maps as the Moine Thrust and the Dalradian are not the only outcomes of continental collision. The closure of the Iapetus not only altered the sediments that were already present, but at the same time set up a new super-continent with new patterns of erosion and deposition. The closure of Iapetus signalled the making of the Old Red Sandstone continent.

A giant continent was built from the components of everything that had gone before and that continent would bring an entirely new climate. Britain, united by the collision of Avalonia and Laurentia, would find itself at the edge of a new supercontinent: Laurussia. This would, in turn, over the course of the next 100 million years, become part of an even larger continent called Pangaea. Britain would be at the centre of the world and at the heart of a vast continent.

CHAPTER ELEVEN

The Devonian Rivieras

Devonian period – from 359 million to 416 million years ago

No Cornishman on earth would ever admit it, but most of Britain's most south-westerly county is Devonian in origin. As we have already seen, except for the Carboniferous granite bosses of the moors, it owes its geological identity to a period named after its arch rival on the other side of the River Tamar. In fact, despite Cornwall being nowhere near as large as its English neighbour, it probably has the same area of Devonian rocks outcrop as Devon itself, so it is surely only an accident, both of history as well as regional identity, that this period was not named the Cornubian[1] by the geologist Adam Sedgwick in 1839 and missed out on the honour of being the only county to have an entire geological system named after it. Had it taken that honour, the Cornubian influence would extend over the whole of Britain as the Devonian does, bookending the country from the far south-west to its limits at the north-east of Scotland.

Once, as a young man studying geography, I leapt to an

1. Perhaps it should be: you write to the International Commission on Stratigraphy about it and I'll take a nice long nap.

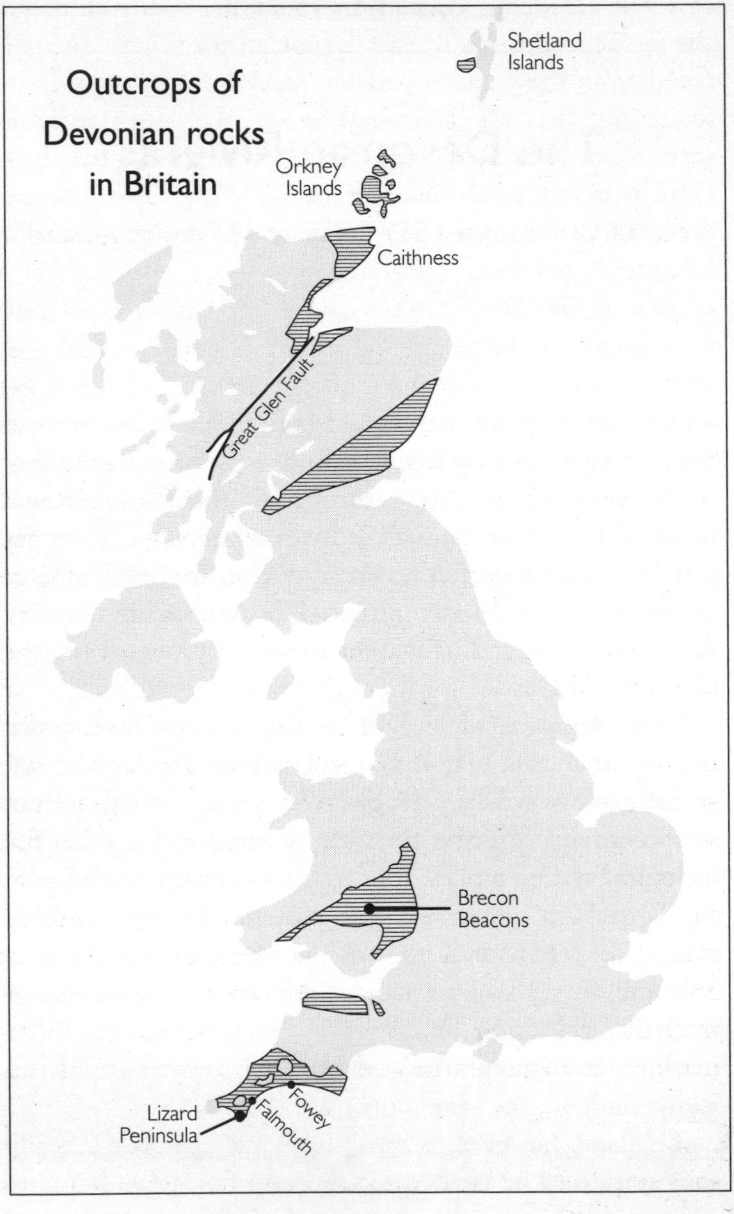
Outcrops of
Devonian rocks
in Britain
Shetland
Islands
Orkney
Islands
Caithness
Great Glen Fault
Brecon
Beacons
Fowey
Falmouth
Lizard
Peninsula

easy but erroneous conclusion about the sandy cliffs and the fertile, red fields of east Devon, scenes that I remembered from the caravan park holidays of childhood. I had found out that the Devonian period had a nickname of sorts – the Old Red Sandstone – and, given that those cliffs were red, sandstone, definitely old and incontrovertibly in Devon, I made the leap. As we have already seen in Chapter 8, I was completely wrong.

During the Devonian period, South Devon and Cornwall were not capable of producing sandstone because they were marine environments and the bottom of the ocean usually has no sand: the streams and rivers that carry it as their cargo have long left it behind in estuaries, deltas and the shallow waters close to the coast. Sandstone instead is formed in three particular environments: in rivers and torrential floods over otherwise dry land like the Torridonian of the North West Highlands; in deserts like the New Red Sandstone; and in shallow seas fed by vast river deltas like the Old Red.

The Devonian represented a step change from earlier periods of marine life with the rapid proliferation of fish species, one of which – the placoderm – may have invented sex according to recent research. A fossil of a smaller fish contained within a placoderm was previously believed to be the larger fish's last meal, but palaeontologists have now started to believe it is the fossil of an embryo which, at 380 million years old, makes it the oldest consequence of internal fertilisation found yet in the fossil record. Other marine developments include the first appearance of that iconic mollusc, the ammonite.

On land however, a revolution was underway. Stable soils started to be created for the very first time and moss forests, which at the outset of the Devonian had been present for a few million years, were joined by plants that became rapidly more sophisticated as time went on. By the end of the period, leaves, roots, seeds and woody tissue were all present and the forests had formed. Hiding in the undergrowth, insects and other arthropods were quick to take advantage of the new habitats.

Between land and sea, the first four-legged animals and ancestors of the first amphibians, the lobe-finned fish, made an appearance towards the end of the period also. The Devonian in Britain is something of an amphibious affair, particularly with regard to the depositional environments found in the far south-west.

In very broad terms, at least, the Cornish Devonian is really just a westwards extension of the rocks of the South Hams and, despite Cornwall's unique qualities, of which there are many, there are perhaps more similarities between the south coasts of the two counties than is freely admitted on either side of the Tamar. This similarity is nowhere more eloquently expressed than in the tourism industry, where the mild, subtropical south coast of Cornwall and the south and east coasts of Devon were both seized upon by marketing professionals from the 1960s onwards and changed by sleight of hand into the English and Cornish Rivieras. In a clear development of the Regency and Georgian fixation with the seaside, tourism chiefs looked to the Mediterranean in an attempt to give the south coasts of Devon and Cornwall the same aspirational garnish as the Côte d'Azur. The notion of a wide, safe bay with rockpools to one side and a white sandy beach underfoot is a powerful lure for tourism – indeed, with added sunshine, it is all any of us want for a week or so. The Devonian rivieras, if I may group them thus, offer an embarrassment of such riches. Strung along the south coast from headland to headland in great swags like celebratory bunting, the coves and bays of the south-west owe their form to the way that the Devonian geology has been treated since it was laid down over 400 million years ago.

Sitting in one such bay at Bigbury-on-Sea, just east of Plymouth one late spring day, I was, in my daddish way, attempting to turn my two-year-old daughter's sandcastle into a major engineering project when my attention was drawn away from the beach and onto some nearby cliffs. A large cloud had broken loose from its pack and a shaft of low afternoon light was striking the cliffs. They reflected the sunlight sharply enough to burn a good after-image of themselves onto my retina after just a glance or two and it seemed almost as if the cliff itself was incandescent – bright white with the acquired sheen that only an advertisement for washing powder or toothpaste can aspire to.

A little later, I picked up a pebble from the bottom of a rockpool that my daughter, having grown tired of my evolving motte and bailey, had moved on to explore. The pebble, though a warm pinky-brown, appeared to have the same lustre as the cliffs and a slaty cleavage.[2] It was one of the most beautiful slates I have ever seen, although it wasn't exactly a slate you would use on your roof – it didn't split that evenly and seemed far too lustrous and handsome for such a common or garden use. Some slates, like the

2. In geology, cleavage refers to the way that a particular rock splits along a line of weakness. Slate, for example, splits in such a way that makes it ideal for use as a roofing material.

famous Llanberis of North Wales or the Delabole of Cornwall, are a just-so assemblage of circumstances; this shiny specimen seemed to be something of a near-miss, a beautiful failure.

I was told once by a geologist friend not to worry about identifying anything tricky that comes from the cliffs on Bigbury Bay, as it is a notorious hotch-potch of partly metamorphosed siltstones, mudstones and slates, folded and tilted with accompanying igneous intrusions. His advice highlighted a key part of what it is about the Devonian riviera that gives it its distinctive coastline. Many of the slates and slate-ish rocks lie with their cleavages in vertical or near vertical alignments and in some places, if you forgive the expression, they expose their cleavage to the elements. The sea seeks out any weaknesses in the slates exposed in this way and erodes selectively to form bays, while those bands of rocks that are slightly harder, or are aligned in a slightly different direction, form the headlands.

Some of the coastline's features, however, appear to happen as a result of factors completely unrelated to the rock's structure and more to circumstances on a regional level. From Berry Head, the southern headland that marks the topographical edge of Tor Bay in Devon to the Helford River in Cornwall, the coast takes on something of a pattern, a basic plan that is repeated over and over again.

The recurring theme of this coast is of high, rounded cliffs of slate and siltstone arranged in a series of bays interrupted by yawning rivermouths so deep and expansive they are open sea in all but name. In geographical parlance they are known as 'rias', flooded river valleys that have existed since the end of the last period of glaciation, the Devensian,[3] which affected Britain from about 115,000 to 10,000 years ago. Although Cornwall and Devon were too far south to suffer glaciers, when the Devensian came to an end, even unglaciated areas were affected as global sea levels rose around 400 feet (120 metres), a state of affairs which had a profound effect on the entire south coast of Britain. Coastal valleys were inundated by the sea from as far afield as Pagham, Chichester, Langstone and Portsmouth harbours through Southampton Water and the Solent, in the east to Poole Harbour and the estuaries of the rivers Axe, Exe, Teign, Dart, Yealm, Tamar, Fowey, Fal and Helford. For the first part of the thaw, sea level rose a centimetre a year, or one metre a century. Once it slowed to a tenth of that, the sediment carried by the rivers overcame the rising tide sufficiently to create estuarine mud banks further up the river.

While there is often little in the way of estuarine muds at the river mouths of the rias – the river has already met the sea miles back in the hinterland – there is usually a sheltered harbour which has sometimes developed into an industrial port on the coast. The very deepness of rias attracts the construction of ports and harbours even when road and rail communications are poor, as is often the case in the south-west of England.

Fowey in Cornwall is one such case. In the heart of Daphne du Maurier country, a pretty little town has grown up along the shoreline of the River Fowey and, like many such places on the Devonian Riviera, it clings to terraces built on vertiginous slopes. Fowey (pronounced Foy, if you

---

3. Not to be confused with the much older geological period (and the subject of this chapter) the Devonian.

don't wish to be corrected when you visit) is hemmed in to the west by steep hills and grew up in spite of difficult lines of communication inland, because the deep water channel of its drowned valley allowed for the export of locally quarried china clay in vast quantities. So essentially isolated is Fowey that moving industrial cargoes by road today is almost as difficult as it ever has been. What was once a 4 mile (6.5 kilometre) long branch line from the little inland village and mainline railway junction of Par has subsequently been converted into a private road by the company that operate the quarries to move material to the clay port, which is upstream a mile or so from the town. The mere presence of the road, which passes through a mile-long tunnel, illustrates how difficult transport can be in such narrow valleys, but also the economic imperatives that a flooded river valley can generate.

Twenty-five miles (40 kilometres) to the west, Falmouth sits at the entrance to one of the largest of all the Devonian flooded valleys, beaten in extent only by Plymouth. Falmouth is a large port, as well as a dockyard, a yacht haven and a centre of all things nautical. Its drowned valley makes it the world's third deepest natural harbour, after Rio de Janeiro and Sydney, and its size and suitability for use as an anchorage has been a key part of its development. First noted by Sir Walter Raleigh who made representations at court and urged his friend Sir John Killigrew to develop the town, Falmouth grew from a few inconsequential hamlets into a bustling town, quickly dwarfing its ancient upstream neighbour Penryn which had a 400-year head start on it. From 1688, the famous packet ships plied their trade from the port and cemented the fortunes of the town, placing it at the centre of the Old World's postal system. Despite being even further west than Fowey, the safe anchorage of Falmouth's ria – the Carrick Roads – made it more than suitable for a large fleet of the world's first dedicated mail ships. It wasn't until 1850, with a new railway to connect it to London, that the port of Southampton took over the trade.

The development of the area continued, however; not much more than a century or so ago, attention was focused on the tiny village of St Just in Roseland, a little upstream of Falmouth and on the opposite bank. It was claimed that there was 14 fathoms of water and enough space for the entire Royal Navy of the time. A few years later, in 1919, there was even an Act of Parliament passed to start building a railway from St Just to a station east of Truro on the mainline to facilitate the scheme, but money problems stopped any further progress. The little church and graveyard of St Just is set in an utterly beautiful mature sub-tropical garden, situated on a green quay which is drained every tide by a Cornish creek, a spot that is the nearest thing to heaven on earth. Anyone lucky enough to visit should be grateful that for once money or, rather, the lack of it, stopped something terrible happening. Today, St Just in Roseland remains a tiny village, with only a few words in every tourist guide to alert you of its presence.

While commercial ports like Fowey and Falmouth developed at the coastal end of their rias, the lowest bridging points of these wide inlets was often miles inland around a prosperous little town or city that grew up as an inland port – sheltered not only from the weather, but also from piracy and the ships of enemy nations – and which thrives to this day as a local centre of trade and administration. Towns like Totnes, Kingsbridge and Truro lie at the inland end of their respective rias, but further afield from the Devonian their grown-up forms can be seen at Exeter and Southampton.

While the Devonian interior is basically fairly undistinguished pasture, the buxom hills add interest as does the enduring human fascination with bodies of water – these tranquil, flooded valleys are havens of the picturesque, from the meanders where the estuaries lie, to the deeper channels on the seaward end. Upstream, as at St Just in Roseland, the lower slopes of the hills are often thickly covered in native trees which grow all the way down to the water's edge, with the occasional bottomless boat pulled up onto the shore under a leafy bough to slowly melt back into the water over decades. Taking a trip up the higher reaches of one of these magnificent inland waterways always brings to mind *Swallows and Amazons*, the soft plough of the boat's wake laps against the shore and the mind slows and allows for moments of subdued reflection where one's inner landscape comes quietly to a rest.

Despite the peaceful atmosphere of those rias, the rocks that bound them saw great violence almost as soon as they were laid down. During the Devonian, the south-west started to feel the outer ripples of movements further afield, of other encroachments responsible for the creation of the Pyrenees and the Urals. This was the Variscan Orogeny[4] and although it did not build great chains under the British

4. The Variscan Orogeny is named after a Germanic tribe, the Varisci who once lived in Saxony. The term 'Variscan' originally had a highly specific meaning that distinguished it from the word 'Hercynian', but the two words are now synonymous, Variscan being more popular.

Isles like the Caledonians it nevertheless had an enormous impact on the landscape as we see it now. We have already referred to it during the Carboniferous period (around 359 million years ago) and it finished in the Permian (280 million years ago) by creating the granite of the West Country moors, but two areas of the south coast saw the opening shots of the Variscan, right at the end of the Devonian.

Science hasn't yet fully settled on an age for the Lizard peninsula which is at the southern tip of Cornwall and, therefore, mainland Britain. It used to be believed that the Lizard – and, by extension, its companion headlands at Dodman Point and Start Point – were Pre-Cambrian in age but there is a consensus building that it might be early Devonian at around 400 million years old.

For all of that, it certainly feels Pre-Cambrian and that might have a lot to do with the fact that it is predominately made out of hard metamorphic rocks of the type we associate with the truly ancient landscapes of the North West Highlands. A closer look at the rocks and the tectonic evidence reveals that they are a pick 'n' mix selection of bits of oceanic crust thrust up or obducted over the Devonian continental rocks of the Lizard, Dodman and Start Points. Among the selection is an extraordinary rock called serpentine, a rock so called because a green variety of it has the mottled, spotted appearance of the skin of a snake.[5] Serpentine is easily cut and polished to decorative effect and you may well come across a number

---

5. Just as in zoological taxonomy, where a snake is no more related to a lizard than I am to a golden crown fruit bat, there is also no etymological connection between serpentine and the Lizard. Lizard, in old Cornish, means 'high court'.

of little lean-to shacks in Lizard Town selling knick-knacks carved out of the rock.

We have Queen Victoria to thank for the shacks as she became rather partial to the stone's looks which were well matched to the Victorian sub-funereal style. On a Royal cruise around the British Isles in 1846, Her Royal Highness and Prince Albert visited the area and were impressed by the rock's appearance. Their patronage kick-started a vibrant industry of carving for architectural and monumental masonry – the appearance of the rock is particularly suited to the rather gothic predilections of Victorian Britain.

Serpentine is a rock unlike most others in that it is not formed in the earth's crust but in part of the upper mantle, which is a thick layer of solid and molten rock that starts immediately below the crust – about 20 miles (32 kilometres) under the Lizard – and goes on for another 410 miles (660 kilometres) towards the centre. Hot rock rises within this layer and starts to move horizontally and then falls again as it gets nearer to the surface and cools. This movement creates a conveyor belt effect that moves the earth's crust along, which in turn is the mechanism for sea-floor spreading. The process is perhaps best understood as being like an enormous lava lamp. The Lizard's serpentine is thought to be the lower part of what is known as an ophiolite, effectively a cross-section of the ocean crust and the upper mantle, thrust up during tectonic activity over continental crust. The presence of oceanic crust and upper mantle in a continental environment is a comparatively rare occurrence indeed; normally, oceanic crust, which is denser, subducts – sinks and slides under – continental crust. In the case of the Lizard, Start Point and Dodman Point, the oceanic crust has obducted over the continental Devonian crust, giving us an opportunity to study the geology of the sea floor without getting our feet wet.

The character of the Lizard peninsula reflects the variety of rocks contained in the ophiolite's cross-section of the ocean floor and how they blend in with the indigenous Devonian rocks. To the north of Mullion on the west coast of the peninsula, the landscape is soft and undulating and typical of the rivieran hinterland elsewhere in the West Country. Mullion even looks like a village from mid-Devon, complete with thatched cottages, including the village pub, the Old Inn. To the east and south, things become more wild and ragged as the complex ophiolite makes its presence felt. From the north cliff of Mullion Cove, to the south-west of the village, it is possible to take in much of the varied geology of the Lizard in one view. Below, to the right of the harbour wall, lies the conical stack of Scovarn and on the other side of the Cove is Mullion Cliff – both made of serpentine. Beyond the headland that marks the end of Mullion Cliff, The Vro, the point of Mên-te-Heul is formed of hornblende schist, as is the headland you are standing on. Out at sea lies Mullion Island, made entirely from a pile of pillow lavas which first oozed out of the ocean floor as a result of ancient sea-floor spreading. All of this was formed and then thrust over the native bedrock within a geologically compact period of only around 35 million years.

Rare things are attracted to this exotic geology. In Britain, the Cornish heath – *Erica vagans*, the county flower of Cornwall – only grows on the Lizard peninsula, on the low plateau known as Goonhilly Downs. Heathers usually favour acidic soils, but the Cornish heath, unusually, only grows on alkaline earth – a characteristic passed on by the serpentine rocks of the Lizard. Goonhilly Downs has much of the moorland about it, if you characterise moorland by its wild and windswept nature. Although there are woods in the valleys, there is little tree cover on top of the downs except for the odd commercial plantation with its geometrically precise rows of conifers, seemingly placed with the aid of alien technology.

The plateau has its own technological advantages as well. In 1901, Marconi broadcast the first transatlantic radio signals from the Devonian cliffs at Poldhu Point, to be received at St John's, Newfoundland. When the cliff was formed, St John's and Poldhu would have been a lot nearer and it would not have seemed such a miraculous feat: according to the writers of one Cornish guidebook, at that time he might have even achieved the same thing by shouting. In keeping with the theme of transatlantic broadcasting, in 1962 the communications satellite Telstar 1 was launched from Cape Canaveral in an experiment to relay phone calls and live television from one side of the Atlantic to the other. Unlike modern communications satellites, which have geostationary orbits that allow ground antennae to be pointed at a fixed part of the sky – much like a TV satellite dish of the kind mounted on millions of homes in Britain – Telstar had to be tracked through the sky by ground antennae as it orbited. Goonhilly Downs was picked as the site to build an antenna to transmit and receive the signals partly on account of the topography, as it has wide horizons which maximises the line of sight between ground and satellite. Because of these advantages, Goonhilly quickly became the largest satellite earth station in the world, playing an important role in the televising of world events like the Olympics, the 1969 Moon landings and Live Aid in 1985. Though recently retired as an earth station and taking its place among all the other Cornish tourist attractions as 'Futureworld', there are still over sixty massive dishes on the site, including one – nicknamed 'Arthur' – that boasts a Grade II listing.

Aside from the occasional band of limestone in Devon, much of the Devonian landscape in Devon and Cornwall is made up of slate of varying quality – the Cornish tin miners called it 'killas'. The highest quality material comes from North Cornwall, where, at 425 feet (129 metres) deep and 1.5 miles (2.5 kilometres) in circumference, the famous Delabole Slate Quarry once had the dubious honour of being the largest hole in the ground in the world. It is also the oldest working slate quarry in England, having been mined for a thousand years, which is why much of the county sleeps at night under a roof extracted from it. The landscape here is rather flat inland and it feels like frontier country, like a damp, warm maritime-climate version of the Wild West. Walking or driving around this plateau, grey houses rear out of the horizon in conspicuously isolated groups that seem to scream 'here is human habitation' at the traveller. Everything appears to be made of slate; not only the roofs, but also the field walls, which are constructed in a herringbone fashion locally called 'curzy way', while house walls are often hung with it as extra protection against the elements. The roofs often show off a particular Cornish technique of scantle slating, where the size of slates is graded from top to bottom, being larger at the eaves and growing progressively smaller towards the ridge.

At one time, using local materials, sometimes in an

imaginative and resourceful fashion, was the economic option and importing would have been madness – though Cornwall only offers a selection of granite, killas stone and roofing slate, they are all suitable and highly durable for building. It says something of our world when new developments trying to fit in sensitively with the local vernacular style go to great expense to import materials to mimic those lying under their very foundations. That caveat aside, here, as in other parts of Cornwall, the geology of the area is reflected directly in the buildings.

At the other end of Britain, in Caithness, much the same could be said for the extremely versatile group of Devonian sandstones found there. In 1914, the British Geological Survey published its geological memoir of Caithness. In it, field geologists C.B. Crampton and R.G. Carruthers noted that the local flagstone – a Mid-Devonian Sandstone – seemed to be used for everything:

> Since the advent of civilisation, scattered houses, many villages, and two large towns have been built entirely with rocks of local origin. In the towns and throughout the county the houses have been built, the roofs slated, the roads paved, the fields fenced, and the drains lined with Caithness flags, so varied are the uses to which the rocks are adapted.

Despite the functional symmetry at work here, the rocks of Caithness and the Orkneys are poles apart from those of the Devonian West Country, being derived from different materials and deposited in different environments; the bookends of Britain, it turns out, are united in terms of the approximate time of their deposition and nothing else.

In Scotland, northern England and mid-Wales, the

erosion of a new mountain range built when the Iapetus Ocean closed, the Caledonians, started as soon as the mountains were formed and continued apace throughout the Devonian period, giving rise to whole regions full of the characteristic rock of the time. In this way, the Caledonian Mountains gave birth to an extraordinary group of rocks, a group that is often seen as the signature dish of the Devonian: the Old Red Sandstone.

Thinking about the term 'Old Red Sandstone', I can't help but remember a television advertising campaign for 'Red Rock Cider' which ran during 1990 and featured the strapline 'It's not red and there's no rocks in it'. It pops into my head because the 'Old Red Sandstone' is a convenient tag that describes a broad group of rocks, not all of which are red or – indeed, sandstone – but which were all created from the erosion of the Caledonian Mountains. As we will see, the Old Red wasn't simply confined to the Devonian either but started in the far north much earlier in the Silurian and, in some areas of the continent, carried on into the Carboniferous. It is a group that has its roots not in the abstracted notions of geological time periods but owes its existence to a single defining event. If you live on the Old Red Sandstone, you are living on a mountain, reduced first to rubble, then progressively finer material which was deposited on flat terrain as a dune or a ripple on a river or lake bed.

You couldn't imagine a more different scene from the verdant valleys of Cornwall and the South Hams – even the rather bleak outlook around Delabole – than the rather austere suroundings of the far north-east of Scotland. The contrast they offer to the mountainous region to the west, the North West Highlands, is just as striking however, because although lofty in places, the landscapes of Caithness are primarily flat. They are geologically distinctive also: in an area where crushed rocks and turbulence seem to be all around, the Old Red of Caithness is completely undisturbed; there are no folds, thrusts, nappes or buckling of any kind in evidence over this part of Scotland. It is almost as if the placid environment of their deposition, in a large body of shallow water called Lake Orcadia, has spilled over into their subsequent history.

It would be wrong to typify the Old Red in Caithness, where it manifests itself as the flat sandstone, as somehow run-of-the-mill or humdrum in any way, however. The uplifted plateau that forms Caithness and the Orkneys is halted at great cliffs that plunge down to sea level on the shores of the Scottish mainland and beyond. The perfectly horizontal sandstone has perfectly vertical joints and it is these weaknesses that, when breached, cause the rock to form vertiginous cliffs cut by narrow gullies or deep clefts that are known locally as geos. Eventually erosion leads to the formation of stacks like the 450 feet (137 metre) high pillar of the Old Man of Hoy, famous the world over for the celebrated difficulty of its climb and the lovable eccentricity of its climbers.

Arguably even more famous than the Old Man of Hoy, especially in the wider consciousness, is John o'Groats, the small village which, allegedly, is the most northerly point on mainland Britain. Anyone with a road atlas and the power of sight can see for themselves that this is wrong: Dunnet Head is a few miles further north and the best that John o'Groats can claim is that it is the most north-easterly village on an A-road in Britain, which is hardly likely to make it onto even the most desperate town council's headed notepaper. The village has one of those fingerpost signs that you pay to have your photo taken in front of – like Land's End – and indeed both signposts are owned by the same company, marking the start or the end of what claims to be the longest journey to or from Cornwall.

There is, however, a longer journey. Take the single-track minor road east from John o'Groats to Duncansby Head, adding at least another 2 miles (3 kilometres) to your end-to-end travels, and from where you will have fine views of the village on the A99. More importantly, you will have fine views of the Devonian landscape at its very best and it will be a lot less crowded. There are views across the Pentland Firth of Orkney Mainland and South Ronaldsay, of Stroma, Swona, Hoy and the Pentland Skerries, but the highlight is a mile south down the coast: a view of the Thirle Door and the Stacks of Duncansby. The stacks, in particular, are worth a journey of any length to see, so the extra 2 miles from John o'Groats will repay every single yard. These two great stacks lie just off the coast as if they are the trailing wings of some huge, supernatural bird or the tailplane of a rock spaceship that didn't quite make it to land. The stacks don't have the 'pipe on its end' profile of the Old Man of Hoy but, rather, are roughly conical and, bizarrely, appear from some angles to be almost a matching pair.

From John o'Groats the rocks of the Scottish Devonian continue in an ever-thinner sliver down to Inverness, where something quite extraordinary can be seen. You don't have to go all the way to Inverness to see it, however, as it is also perfectly visible from space and, therefore, from online satellite photographs; it is the Great Glen Fault.

In terms of British fault lines, the Great Glen Fault is the big one. On land it runs for over 90 miles (150 kilometres) from Inverness south-west along Loch Ness, Loch Oich and Loch Lochy, through Fort William and out towards the Inner Hebrides via Loch Linnhe. It continues on under the sea from both ends: to the north-east it marks the eastern coast of Scotland and probably extends from Caithness towards Shetland; to the south-west it continues through north-west Ireland through Lough Foyle and Donegal Bay. There is even strong evidence that links the Great Glen with the Cabot Fault in Newfoundland, which line up precisely with it, once the 2,000 miles of the Atlantic Ocean is removed from the equation.[6]

The Great Glen Fault imposes its form on the landscape in an obvious way – the Great Glen is itself a product of the fault. Even without its international affiliations, it is easily the most substantial and destructive fault structure present in the British Isles and, although its work is done for now, the occasional moderate earth tremor hints that there are still small amounts of energy being stored and occasionally being released. The presence of earth tremors inevitably invites comparisons to the most famous fault of all, the San Andreas in California, responsible for some of the most devastating earthquakes of the twentieth century. The Great Glen is a similar structure, a tear fault, with the landscape to its north moving approximately 80 miles (128 kilometres) to the south-west during the Devonian, which was then followed by a movement back of approximately

---

6. The Atlantic Ocean started to open up only around 130 million years ago and, therefore, did not exist during the Devonian.

18 miles (28 kilometres). This sideways movement was responsible for the formation of a belt of shattered rock half a mile wide and provided an easy path for glaciers to gouge away at. Loch Ness, which is an average of 600 feet (182 metres) and in some parts 750 feet (228 metres) deep, owes that depth to the glaciers that followed to carve out the weakened trench. You can still see some of the shattered rock in a cutting by the road at Castle Urquhart on the north side of the water. You'll see it a long time before you spot anything stirring in the Loch, that's for sure.

Nessie was supposedly left behind, stranded after the Devensian ice melted 10,000 years ago, but there are stranger (and verifiable) marine offerings in the Devonian rocks of Scotland. It is perhaps appropriate that the Devonian, named after a county so rich in maritime heritage and one that once carried as its badge a simplified Elizabethan-era sailing ship set on the swash of a wave, should also be the age of fishes, albeit ones with an ability to exist in freshwater environments. Fish are everywhere and nowhere in the Devonian. The Old Red does not make for an easy palaeontological hunting ground, but an experienced fossil collector can unearth fantastic specimens, particularly back up in Caithness.

One of the most famous spots in Scottish geology, the Achanarras Quarry can be found in the north of the old county. Lying near the A9 between Latheron and Thurso, the unfortunately named village of Spittal is the closest settlement and you can follow directions to Achanarras from there. Formerly a quarry for extracting roofing slates, Scottish Natural Heritage have done a fantastic job in opening up the site and providing context boards in an information hut, but since its heyday in the nineteenth century the site has been rather over-collected and the previously abundant supply of whole fossils has dried up, though there's still plenty of remains. All of the quarry's very well-preserved fish, from fifteen different genera, are thought to have been deposited in just 4,000 years.

The Old Red also outcrops further south in Scotland in the Midland Valley, another south-west–north-east fault-bounded area which stretches from coast to coast between the Highland Boundary and Southern Upland Faults. One particularly interesting freshwater animal was found near Arbroath, a fishing village on the east coast of Angus to the north-east of Dundee famous for its delicious smoked haddock 'smokies'. I'm sure all sorts of things have been trawled up in the nets offshore, but this was something extraordinary: a *Pterygotus* which is nicknamed the giant sea scorpion. With the strapline for Red Rock Cider still ringing in my ears, I am delighted to inform you that the giant sea scorpion is an arthropod that did not live in the sea and was not a scorpion, although it does bear a passing resemblance to one. It was, however, definitively a giant at 6 feet (2 metres) in length and one of the largest arthropods ever found.

So far in this account, the Old Red has given rise to rather muted landscapes. It is, as anyone who lives in a house made of it knows, a comparatively soft rock and so tends to erode rapidly and generally forms rather flat landscapes. In certain circumstances, however, it can endure and one such durable outcrop occurs where we now turn our attention: the Old Rcd Sandstone's most grand expression, that of the Brecon Beacons. The Beacons are part of a large triangular formation of the Old Red with its apex near Bridgnorth in Shropshire, its thin western end in Pembrokeshire and its eastern edge shadowing the banks of the River Severn down to its estuary. In South Wales, the base of the triangle is hidden under a large area of Carboniferous coal and later rocks.

There is something elemental about the Brecon Beacons, the western half of which, the Fforest Fawr, has been designated a European and UNESCO Geopark in recognition of its geological charms. The Beacons have severe and precipitous north-facing scarp slopes and giant cwms plucked out by glaciers all of which undoubtedly add to their raw menace. In places, the scarps join up to form chains and one such chain joins the northern scarps from near Ammanford, north of Swansea, via the Black Mountain, the Beacons and the Black Mountains[7] all the way to the English border – a stretch of 50 miles (80 kilometres).

Halfway along this scarp, it seems that Pen y Fan – which is also the highest summit in Britain south of Snowdonia – is set to remain durable for a long time because of the massive[8] sandstone called the Plateau Beds at its peak. Together with the linked summit of Corn Du, the Plateau Bed of Pen y Fan protects the underlying sandstones from the elements. The Plateau Beds are only one thin bed, around 180 feet (60 metres) deep out of an Old Red depth that runs into the thousands of metres. But here they control the landscape, protecting the softer beds

---

7. Confusingly, the Brecon Beacons are flanked on the west by a range known as the Black Mountain and to the east by another range known as the Black Mountains.

8. Massive, in a geological sense, means it has no discernible form or structure, that it is, simply put, one piece. As form and structure is shown by planes and joints – lines of weakness – a massive structure is likely to be stronger.

below, dictating the lie of the land. Further south, all the Devonian beds disappear, hidden under the enormous South Wales coal fields, as well as its associated deposits of millstone grit and limestone.

CHAPTER TWELVE

The Long March to Siluria

Silurian period – from 416 million to 443 million years ago

Borders can be dull places sometimes and there is often precious little on the ground that supports the potential of the red line on the map. It seems almost naive to say it now in the all-knowing international wisdom of our times, where a flight to the Mediterranean usually costs less than a train to Colwyn Bay, but I remember that, when I was eight, a trip to Kent in a Hillman Avenger was a voyage to foreign climes laced with adventure, particularly as we were going to somewhere called Royal Tunbridge Wells. Now, of course, my inner cynic squirms at the very thought of it, but I vividly recall how, on that journey, I saw for the first time how shafts of sunlight radiated from behind a fair-weather cloud, spotlighting patches of woodland and ripening corn on a hillside in the middle distance. I think I must have got counties and countries hopelessly mixed up, because we lived in Sussex at the time, but for a few years after, I believed that this amazing phenomenon only happened in Kent, so sold was I on the idea of boundaries marking a gateway to the exotic.

Now, in my mid-forties, things are different, but the landscape of the Welsh Marches defies the grown-up view

Outcrops of Silurian rocks in Wales and the Welsh Borders

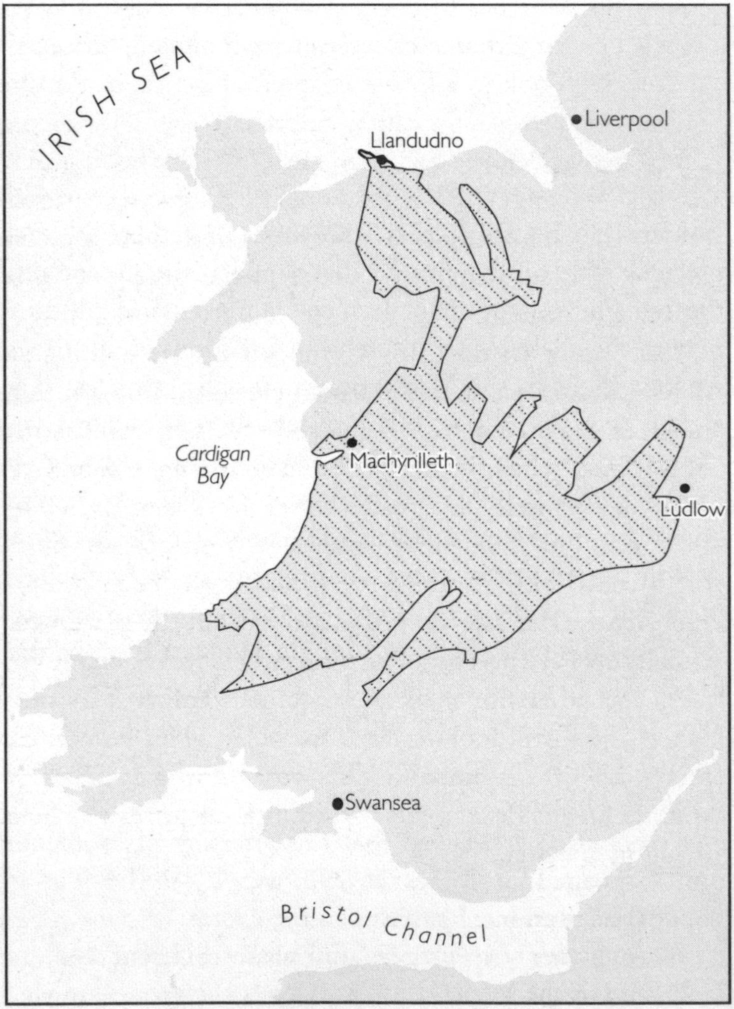

that borders are just dull, arbitrary lines drawn for political convenience. And while I know that the English–Welsh border is not a full passport-packing, visa-stamping international border, there is, nevertheless, a physical dimension to it, particularly if you accept that the Principality overcomes the Marches to create a Greater Geological Wales. Much of that extra territory can be found in Shropshire and we will look in a future chapter, at depth, at the Pre-Cambrian history of Church Stretton Fault – nicknamed Little Switzerland – but they are not the whole story of Shropshire, merely a small sample of their most magnificent hills. One cannot pass by Shropshire and talk of its hills, of course, without reference to the poetry of Alfred Edward Housman.

According to the people who dissect such things, in *A Shropshire Lad*, A.E. Housman eulogised first and foremost his own emotional landscape rather than that of Shropshire, which is merely the stage. Being a Worcestershire lad himself, he apparently didn't know the county that well, but frequently gazed at it from afar at the top of Bredon Hill. Shropshire would have been west of Housman – the direction of longing itself, the place where we all look at the end of the day when we're feeling lost, alone or otherwise in need of some comfort. Housman, indeed, was still looking west in 1922, when he collected forty-one of his unpublished works for *Last Poems*, a volume for his dear friend Moses Jackson to read as he lay dying in Canada.

When summer's end is nighing
And skies at evening cloud. I muse on change and
  fortune

And all the feats I vowed
When I was young and proud.
The weathercock and sunset
Would lose the slanted ray. And I would climb the
beacon
That looked to Wales away
And saw the last of day.

Housman's connection with the 'blue remembered hills' of Shropshire may have been more literary than literal (he wrote the cycle of poems in London, after all, and only visited some of the locations after the book was published), but Shropshire Tourism make the most of it anyway. And why not? They can do this and get away with it because Housman painted a memorably simple and effective portrait of Shropshire as an English rural idyll and because the Shropshire Hills measure up in every sense with Housman's poetry which is both plain and romantic at the same time. The Shropshire Hills are, indeed, very special places.

Not least among the hills is Wenlock Edge, a long and lofty escarpment that runs from Craven Arms, 7 miles (11 kilometres) south of Church Stretton on the A49, north-east 16 miles (25.5 kilometres) or so to the town of Much Wenlock. Much Wenlock's white church is the source of its name, which is believed to have come down from the Welsh Gwyn-loc, a 'white place', referring to its white limestone. This is the same Silurian white limestone that makes up Wenlock Edge. The edge itself dips down to Hope Dale and then another, younger limestone escarpment to the east rises to the much higher Aymestrey escarpment which, in turn, dips to Corve Dale. These repeated ripples of scarp, dip and dale are a result of the alternating limestone and shale geology of the area and the subsequent Variscan[1] folding which led to the creation of an anticline long since eroded down to the escarpment landscape we see today. Each hard limestone escarpment overlies a softer shale valley and because, broadly, shale is deposited in deep-water environments and limestone in shallower seas, this alternating deposition is an effect of repeated regression and transgression of the sea.

The Much Wenlock limestone of Wenlock Edge is well known for its fossilised coral reefs, formed in warm shallow seas (less than 100 feet or so of water) when the area was 25 degrees south of the equator. A walk along the edge now – perhaps in less than equatorial weather, if mine was anything to go by – will reveal the calcite remains of these ancient reefs, which, in a few spots, are easy to see. If you don't make it up into the hills, a trip around the area will reveal that particular time-worn mildness that vernacular buildings made from limestone always seem to exhibit. Many of the houses in the villages and towns around Much Wenlock are either made from the limestone or are of the cruck-frame half-timbered variety – the combination of the two lending the towns an antique friendliness unsurpassed anywhere north of the Cotswolds.

Looking at the limestone beds in a spirit of recreational curiosity, either scientifically as fossils or in admiration of their aesthetic qualities as a building material, is a luxury of our times, however. At one time, long before Housman, Shropshire limestone was not so much a thing of beauty as

1. Variscan folding occurred in the Carboniferous period as a result of the final assembly of Pangaea, the last super-continental formation before the break-up into the familiar continents we know today.

an industrial resource. Limestone was a vital ingredient in the Midlands which, along with plentiful nearby coal and iron ore, fuelled the Industrial Revolution that occurred at the same time as the gentleman geologists of the Age of Reason were roaming over the hills and dales of Shropshire in search of fossils. The coal may have been mined as early as Roman times. Beneath the baths of the city of Viriconium,[2] traces of local coal have been found in the central heating system and the Romans would have almost certainly discovered it while they were constructing Watling Street, as a seam outcrops along the route near Oakengates.

Over the course of the eighteenth century, limestone was used in the production of iron, using all the local materials at their disposal as well as the expertise of Abraham Darby and his two sequels, Darby II and Darby III at the Coalbrookdale Ironworks. Although iron had been made around the area for a long time using charcoal that was scarce and expensive, it was not until Abraham Darby I started using coke to make it in 1709 that industrial quantities could be manufactured relatively cheaply. His son, Abraham Darby II, improved the quality of coke-smelted pig iron so that it could be forged into wrought iron and Darby III was responsible for constructing the world's first iron bridge. It united the north and south shores of the River Severn at what later became the village of Ironbridge, near Coalbrookdale, in 1779.

Man had been adapting the landscape to his own ends for thousands of years but in 1779, for the first time, geology had provided the raw materials for an industrially manufactured landscape feature. The graceful lines of the

2. Viriconium is now better known as Wroxeter.

Ironbridge made an impressive sight and became an attraction almost immediately. Eighteenth- and nineteenth-century tourists and painters stopped off on their journeys towards North Wales to witness this newest of landscapes. The most famous painting of the area is not Darby's bridge however, but the fiery image of *Coalbrookdale by Night* by P.J. de Loutherbourg, painted in 1801 and now owned by the Science Museum in London.

The case of Ironbridge, the coal and limestone, the canals and then the railways that followed pretty much place part of the rural county of Shropshire at the hub of eighteenth-century engineering technology, but by the early half of the nineteenth century, Shropshire was becoming the centre of pure scientific inquiry.

The whole of Shropshire is another of British geology's Meccas and of all those I've visited, I've never encountered such widespread public awareness and pride in the local geological heritage. The county has some of the most varied scenery in Britain, probably due to the fact that, of the whole of geological time, only those three periods that represent the geological record between 145 million and 2 million years ago are absent. Murchison named two of the now internationally recognised epochs of his Silurian system after Ludlow and Wenlock – something that seems to have seeped back into the soil of the area itself, because everyone seems to know something about it. During a lunch break on the gusty peak of High Vinnals in Mortimer Forest in the far south of the county, I had an interesting conversation with a woman who claimed to know nothing about geology, but was interested enough to know a thing or two gleaned from the excellent local museum at Ludlow. We discussed the buzzards, ravens and the pair of goshawks that lived in Mortimer Forest and she showed genuine interest in the trilobite fossils I had just collected from a cutting by a stream, while her hound, according to its pathetic gaze, indicated that he showed an even keener interest in my cheese and pickle sandwiches.

I had come to Mortimer Forest to collect fossils. It was among my first-ever brushes with palaeontology, if you ignore the odd walk along Lyme Regis beach pointing at ammonites the size of a small cartwheel. I had chosen the location carefully from a number that were featured on a website and had travelled to Shropshire to find trilobites, the ancient arthropods that seem to encapsulate the early radiation of life in the seas. Wary of an earlier trip in Wales to find a contorted *Angelina* trilobite, which had me wandering around the wrong part of a hill for an hour, it was a relief when the website and its paint-by-numbers idiots' guide instructions took me straight to the spot where, I was promised, little *Dalmanites myops* may be found in relative abundance. After about five minutes of clawing away at a damp shale embankment, half the time lying on my back in a puddle, I emerged triumphant. As I stood there – one side of my face accidentally daubed with the pale, creamy mud and looking like an asymmetrical, overweight Neolithic warlord, I held a block of stone that had survived for around 420 million years, but all of a sudden seemed to be halfway back to being mud – the edges quickly deteriorated with handling in the same way soap turns to mush in a warm bath. I set it down on the scree under the bank and tried to detect the best place to hit it.

From its edges I could just about work out which way

the bedding planes went and gave it a soft tap on its end. To my utter amazement it split evenly to reveal a little *Dalmanites* tail, less than the size of a penny piece. I was already hooked and soon had found something else, a *Monograptus tumescens* – a graptolite, the distinctive fossil of a small colony of filter-feeding animals, each in its own cup along a common stem. Graptolite means 'writing in rock' and they do indeed bear a passing resemblance to tiny copperplate scribbles but my preferred description is of a section of broken bandsaw blade. Graptolites, which seem to have been widespread, are used extensively by geologists as a zone fossil: their rapid evolution means that it is possible to divide up the Ordovician and Silurian into manageable million-year chunks.

By the time I had walked up a tortuous hill to my lunch, powered more by exhilaration than by any hint of physical fitness, I must have been exhibiting the same degree of obsession with my discoveries as that kind lady's dog had shown for my sandwiches. My bag weighed about eight pounds more than it had done when I arrived and I suspected further Variscan folding had made the slopes even steeper than they appeared on the map, but I'll never forget the moment when I first saw the little *Dalmanites*, freed after 420 million years in the mud.

I walked back to nearby Ludlow. A beautiful town, a community alive with the air of antiquity and old-world refinement and one that was recently described in the glossy pages of an upmarket property and gumboots magazine as the 'most vibrant small town in England'. It has an atmosphere somewhere between a market town like Evesham and a small cathedral city like Bury St Edmunds, which are two and three times as large, respectively.

Ludlow has confidence and gravity way beyond its size and you might idly wonder why that might be. Like a scientist searching for dark matter to account for an unexplained gravitational effect, you could be forgiven for being interested in where all this consequence and magnitude might be originating.

On my way back from Mortimer Forest, I inadvertently discovered the answer. The path down from the forest leads to a minor road where all at once the magnificent Ludlow Castle appears. The castle is a medieval fortress that overlooks the River Teme on the western flank of the town in something of a domineering and forceful manner. Built out of stone from the outset – one of only a handful of castles not to have had an initial wooden phase – it was erected from the end of the eleventh century on, using Silurian siltstone quarried from its own site. Though grand and forbidding on its western side, it is perfectly proportioned on its eastern side – almost hidden – where an entrance opens out onto the fairly small marketplace of Castle Square.

I had initially arrived in Ludlow from the opposite direction from the railway station and hadn't noticed the castle at all. In fact, my first impressions of Ludlow were a long way short of impressive. Over many years the frequent sensation of initial disappointment in the close proximity of luggage has gradually made me realise that first impressions don't count for anything when you arrive by rail. In many towns in Britain, the railway station can be found at the rough end, whereas in others, for historical reasons, it's in the industrial part of town. More often than not, in any case and in a strangely self-referential way, the railway is always on the wrong side of the tracks.

I was immediately and thoroughly overwhelmed by the ennui you suffer when you see yet another branch of Tesco with its gluttonous appetite for parking space and the contents of our wallets. It doesn't matter that they put a roof on it that clearly pretended to allude to the shape of the Shropshire hills behind, it's a building that is quite plainly like all the others. Though even that flatpack monstrosity was better than next door: an Aldi that looked, in the way that a branch of Aldi always looks, as if someone has gutted a car showroom and then quarter-filled it with opened boxes of biscuits.

Roderick Impey Murchison arrived in Ludlow 168 years earlier than I did but you have to give the man credit, he certainly left his mark because they are still talking about him all around town. Unfortunately, as with so much of the conversation around Sir Roderick, some of it isn't particularly flattering, and there is more than the slight whiff of indignation about it all.

Like other parts of Shropshire, the acute awareness of the surrounding hills and of the early geologists who crawled over them in the name of natural philosophy seems to be very deeply ingrained around Ludlow; but the town is one of those rare places that seems as if it was always interested in and, probably, always prosperous enough, to sustain an interest in the natural history of the area. The gentlemen scientists of the town set up one of the world's first natural history societies in 1833, an offshoot of which was the Ludlow Museum – again, amongst the world's first town museums – just under fifty years before London's Natural History Museum opened in South Kensington. This must have seemed at some level, at least, like a local knowledge mine to Murchison when he arrived and, indeed, Sir Roderick harvested a great deal of it in Ludlow, famously from the Reverend Thomas Taylor Lewis, the vicar of Aymestrey, and Dr Lloyd of the town. Lewis, in particular, was disgruntled that Murchison's *The Silurian System* did not contain a single credit to his work or insights, both of which he generously shared. A copy of the letter he wrote to Murchison is held by the Edinburgh University Library.

> I cannot withhold from you that I felt disappointed in the slight notice my early researches have received in this volume. Looking, or I should say, watching as I have the progress of the subject for the last 23 years, I cannot be ignorant of the importance of my early doings, of the accuracy of the succession I had observed of the rocks in the neighbourhood of Aymestrey (the equivalents of the Upper Silurian) previous to your first visit to that locality, and of the value of my subsequent identifications and of the richness of illustrations I there laid before you, and the liberality with which I continued to supply you with every thing that came within my reach, and as you acquiesced in the estimate given of my labours by Dr Fitton in the *Edinburgh Review*, I had flattered myself, as others thought, that whenever you reproduced *The Silurian System*, you would record there a little more detail.

In 1835, Lewis and Lloyd had found what came to be known as the Ludlow Bone Bed and shared their discovery with Murchison. The Bone Bed is a very thin layer of sandstone, 413 million years old, right at the top of the Ludlow epoch of the Silurian that contains small black fish scales and spines. For years, the Ludlow Bone Beds held their place as the oldest fish remains on earth, but Cambrian fish have since been discovered to displace them. What Lewis, Lloyd and Murchison didn't know about the Ludlow Bone Beds was that it contains evidence of something that is at least as interesting as fossilised fish: charcoal. The Bone Beds can now be seen to include the oldest evidence of a wildfire on the planet.

Murchison and Sedgwick had initially set out in the early 1830s to bridge the gap between the ancient rocks of the Pre-Cambrian and the Old Red Sandstone of the Devonian and – politics and foment aside – they succeeded in classifying what were then known as the 'transition rocks' into some kind of context.[3] With the closure of the Iapetus Ocean, the Silurian ended, but such a large collison showed its effects at different times. In the case of the Caledonian folding that followed the Iapetus closure, those effects started in the north and gradually found their way to central Europe dozens of millions of years later. The broad effect at the closure of the Iapetus was to turn mostly marine environments into continental ones. Compression of strata, their folding and uplift turned Scotland dry by the start of the Silurian, but Wales and the Welsh Marches had to wait until nearly the end. The Czech Republic was still wet millions of years into the Devonian period,

---

3. Part of their rivalry was that they were both keen to find the oldest fossiliferous beds and I have to report to you that, eventually, Murchison succeeded – after a fashion – in being associated with ancient life way beyond Sedgwick's capacity to posthumously trump him. In 1969, a 200-pound meteorite fell on Murchison, Victoria in Australia, one of a number of towns named after him. The Murchison Meteorite was subsequently found to contain amino acids and nucleotide bases of the kind found in living cells, RNA and DNA. The meteorite was 4.6 billion-years-old.

a period typified in central and northern Britain by arid conditions comparable with Texas today. In between, the southern part of Britain – particularly the far south-west – would alternate between the two with a lot of deposition in shallow sea shelf and the shoreline being evident. Geologists have a special word for the uniquely geological problem of having the same event occurring at different times in different places: they describe these events as diachronous. So it is that although the boundary – the border if you will – is internationally agreed upon, the geological reality of the Silurian–Devonian frontier is a lot messier and is, in fact, smeared across Britain and, even, the whole of Europe.

CHAPTER THIRTEEN

The Fiery Ring of Wales

Ordovician period – from 443 million to 488 million years ago

Many areas of Britain owe their varied scenery to a complex mix of rocks laid down over hundreds, if not thousands, of millions of years. In places like the North West Highlands it is possible, even easy, to walk over a landscape that spans 2.5 billion years in the making in just a single day. And that's without counting the last 500 million years, where steady erosion, uplift and glaciation have moulded the landscape into its current state. Further south and far away from the ancient basement gneiss, there are still many landscapes that indicate the work of a breathtaking period of hundreds of millions of years, but sometimes there is more than enough interest to be found in a landscape built over a relatively modest time span. One such time is the Ordovician, where 71 million years of volcanic activity, deep and shallow water environments are broadly responsible for a truly remarkable mountain: Yr Wyddfa, or Snowdon.

Although the mountain lies adjacent to the great Cambrian formation of the Llanberis slate which we will deal with in the next chapter, almost everything else on Yr Wyddfa and its surroundings is Ordovician. On my visit

Outcrops of Ordovician rocks in Britain

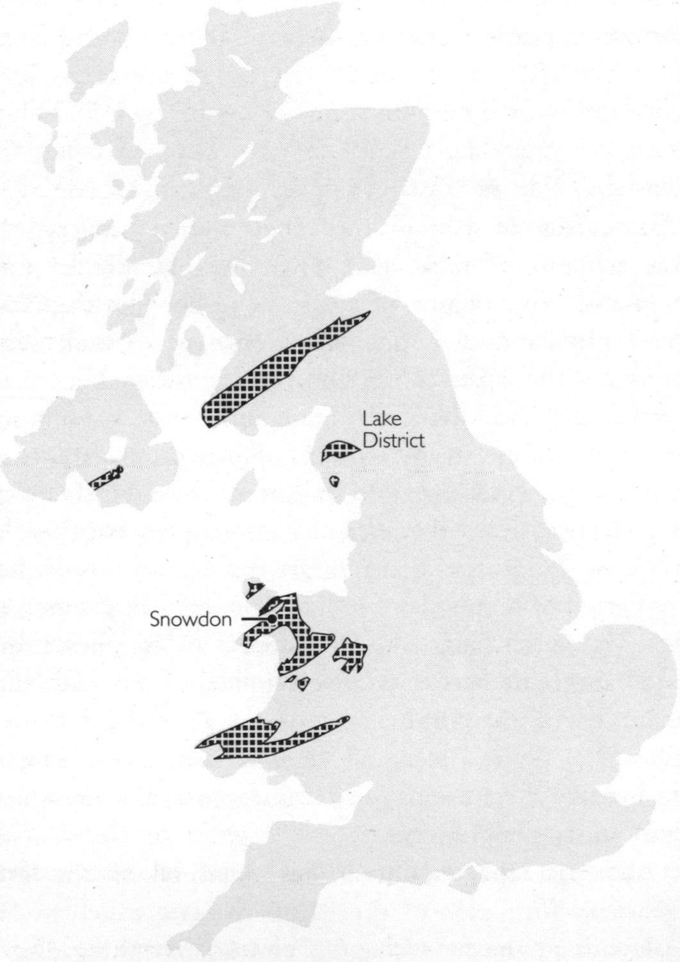

there, one early August day, I came in early on the A498, the road from Beddgelert, and climbed about 600 feet (182 metres) up the Nant Gwynant valley to the site of the Roman camp at Pen-y-Gwyrd, at the turning for the Pass of Llanberis, the A4086. There is a parking spot about two thirds of the way up and the views are phenomenal. Morning cloud was peeling itself off the crags and hung low in the valley below, and a low-angle, numinous light picked out every little bump and hollow of the hills. White streaks of water like tangled threads, plunged down the mountain's side, each with its own torn paper, ragged edge. Perhaps, like me, you always feel at home in the mountains, exhilarated by the air and scenery, and wonder why you would want to live anywhere else. On mornings like that, I wonder why I live on the edge of a flood plain, obscured by a rectilinear hangar of beech trees.

Tearing myself away from the view and my doomed attempt to capture it with a little plastic Japanese box of soulless electronics, I arrived at Pen-y-Gwyrd and turned left up a mercifully short stretch of road, steep rock on the offside while, on the nearside, it seemed to be pure sky. While my mind was drawn to that bit of *The Italian Job* where the coach hangs off an identical precipice, a cloud drifted carelessly across the carriageway of the Llanberis Pass at its highest point – Pen-y-Pass – and I briefly lost sight of the sky and some of the road. The Youth Hostel here is marked on the map as Gorphwysfa, a word which means 'resting place', but which I prefer to think of as the phonetic representation of a cloud rubbing a slate mountain.

Once the potential plunge to a certain death in a Ford Fiesta was safely out of the way, the pass became beautiful once again. I hummed 'We're in the Self-Preservation Society' to myself all the way down to the bottom, stopping briefly at a lay-by halfway down beside the Afon Nant Peris, the tumbling stream that fills the waters of the Llyns Peris and Padarn in the valley.

The premise of this book is based, in part, on the idea that the underlying geology of these islands is reflected in the landscape. But, more often than not, it is in an apparently contradictory manner. We might expect to see a great anticline form a ridge of hills and a syncline underlie a valley and indeed, in parts of the world where the mountains are still being built – some of the foothills of the Himalayas are a good example – you can find smooth mounds that have the rounded contours of a textbook anticline. But once erosion has got to work, nature is usually far subtler than that and all sorts of factors come into play. The hardness, strength and durability of the rock, local tectonic forces and whether an outcrop is porous, permeable or not all play their part and often work to overturn our first expectations. So it is with the second highest mountain in Wales: Yr Wyddfa.

It seems to turn common sense on its head, but the summit of Yr Wyddfa is wholly situated upon a syncline, a great trough in the rippled beds of North Wales, with the highest part of the mountain at its very hinge. You can see the bedding planes mark out a clear V-shape of the syncline – as with all such things, they stand out better in snow – on the east facing sides of the summit. They are best viewed near the top of thc Pyg Track, a more challenging route up the mountain than the longer, easier Llanberis Path and one which is probably best left to an experienced walker, especially when there is snow on the ground. To the south, the limb of a syncline – the Snowdon Syncline – also forms the northern limb of a great anticline known as the Harlech Dome.

Finding the trough of a syncline on a mountain summit only seems contradictory until you actually walk (or take the train) up a mountain yourself. The higher you go, the more you will notice the most important factor in shaping our landscape: the weather. To put it simply, the uplift of a bed of rock adds energy to all the processes of erosion. It opens up the bedrock to more intense wind and rain, a fact easily confirmed by standing on a high hill or mountain. Snowdon's average annual rainfall, for instance, at over 4.5 metres, is almost six times that of the allegedly wet Manchester and winds can often gust at up to 150 miles per hour. Secondly, higher contours means that rain falling on its slopes as well as the water already present in streams and rivers acquires greater potential energy and will cut deeper channels. Thirdly, the very process of curving a bed of rock that was formerly horizontal will compromise its structural integrity to a greater or lesser degree. Finally, we return to the weather; the crest of the Harlech Dome's anticline would have once towered above the present peaks of Wales – some estimates put it at up to 10,000 metres high – and the ice and snow formed at such a height are powerful agents of erosion in many and varied ways, a fact that we have already explored in our chapters on the most recent 'ice age'.

Humans like to think of the patterns of mountains and valleys as durable things but, on a geological timescale, they are only temporary. The natural tendency of the planet's topography is towards a levelled surface where energy potentials and, therefore, erosion will dwindle into insignificance. The only thing that can renew the potential for erosion is further uplift, whether that is the eruption and consequent building of volcanic cones, large-scale vertical movements produced by faulting and further tectonic action of the kind that produce synclines and anticlines.

Snowdon then is an escarpment and all escarpments or cuestas owe their appearance to essentially the same forces and all are the result of the process of differential erosion. Although the Reverend Adam Sedgwick would doubtless approve, I hesitate to seek inspiration for a scientific explanation in the pages of the Bible but, according to the Gospel of Matthew, God 'sends rain on the just and on the unjust' and it's worth noting here that His rain also acts equally on all the rocks in a given area. It is, therefore, the essential qualities of the rocks themselves that will determine how they will weather and how their relative erosion will form the landscape itself.

In the case of escarpments, they are formed when gently tilted harder rock overlies relatively softer material. The hardness of a rock is a product of both its physical and chemical qualities. Shale is inert chemically, but physically weak. Sandstone is made of tough grains of quartz and feldspar – two very hard minerals – but is only as strong as the cement[1] that holds it together. Limestone is very strong in an arid climate (think Egyptian pyramids) but in a warm and humid climate, it eventually dissolves like an Alka-Seltzer. The formation of escarpments relies on relative qualities, however; chalk is a rather soft, even crumbly,

1. In these terms, cement is taken to mean the material that binds the particles together in a sedimentary rock.

rock, but in comparison to the Gault Clay that underlies it in the South Downs it is hard. In all escarpments it is the harder rock which forms the ridges and controls the shape, the dip of its bed becoming the gentler dip slope of the cuesta, while the softer material underneath is eroded more rapidly on the steep, scarp side. Once steepened, streams and rivers start to form gullies on the scarp which, in turn, carry even more erosive power. The scarp slope will then start to erode further back over time.

The hard rock that forms the enduring ridge of the Snowdon escarpment is volcanic lava. The beds of the Snowdon Syncline were laid down during an extraordinary period of time over much of the Ordovician, when the micro-continent on which it was formed, Avalonia, saw intense volcanic activity. Only by the last epoch of the period, the Ashgill which started around 460 million years ago, did this extreme volcanicity start to subside. The Ordovician marks the time that the Iapetus Ocean began to close – when what became the discreet micro-continent of Avalonia rifted away from Gondwanaland, an ancient super continent that straddled the South Pole. Its separation was marked by a 70-million-year fireworks display occasionally more destructive even than the Krakatoan and Mount St Helens volcanoes of the nineteenth and twentieth centuries.

As Avalonia started its 4,000-mile-long journey north, the crust of the Iapetus was subducted under it, pulled down towards the earth's mantle and melted, which led to the formation of a volcanic arc on the surface above the zone of subduction. Avalonia would have resembled modern-day Indonesia but on an even more violent scale. Whereas it is estimated that Krakatoa ejected around

20 cubic kilometres of ash and lava during its 1883 eruption and was heard up to 3,000 miles away, it is believed that one volcano in the Avalonian arc was three times the size, ejecting around 60 cubic kilometres of lava and ash. Some of the remains of all this lava and ash can still be seen on the slopes of Yr Wyddfa.

There are many tourist leaflets and websites that will tell you that 'Snowdon is an extinct volcano', but they are in danger of gilding the lily, so to speak, because while it is no such thing, it is still awe-inspiring. Yr Wyddfa is not an extinct volcano but is near to at least one and may be close to more, the exact locations of which are hard to pin down. It is made, instead, out of bits of those volcanoes, both lava and ash or – to give it its geological name, tuff – deposited on the Ordovician seabed. While nobody knows exactly where the lava and tuff of Yr Wyddfa came from, a pattern of ash and lava formations in Ordovician North Wales shows us that a group of extinct volcanoes encircle the Harlech Dome like a great bass clef, a formation which is sometimes known as the 'fiery ring'. It seems like another Snowdonian contradiction when the majority of Yr Wyddfa was laid down as a sedimentary rock, even though its ultimate source was igneous activity on a grand scale. As if to confirm it, the summit has fossils from the Ordovician sea floor within its matrix. If you get a chance to in the crowds, look carefully at the rock steps of the summit platform. Having Ordovician seabed fossils at the peak of a mountain is a rare quality, but Yr Wyddfa isn't the highest mountain to have them, that honour going to Mount Everest.

The fiery ring of Welsh volcanoes were at the leading edge of Avalonia's trawl north, along with those of the

Lake District. This, along with their similar glacial histories, might explain why the two areas have so much in common – a geological version, perhaps, of the idea of convergent evolution.[2] The subject of the Lake District brings us inevitably to the question of poetry and the poetic appreciation of the landscape, which contained within it the seeds of the curiously British view of the countryside and the tourist industry that started as a direct consequence. While the gentleman geologists of the late eighteenth and early nineteenth centuries were chattering about natural philosophy, artists and writers were immersing themselves in the wonders of the British landscape.[3]

Two fields in particular – science and literature – are often characterised as being poles apart, a view perhaps more informed by modern sensibilities than by any objective evidence. Those views place science in a materialist camp and all artistic endeavour as entirely in opposition, a

2. Naturalists often identify pairs of animal species that are completely unrelated but which are essentially subjected to the same environmental pressures as one another and it says something about nature's dogged pursuit of good design that the process of natural selection often conspires to make them look remarkably similar, their form following their function. In the bird world, for example, swifts and swallows are totally unrelated, but insect feeding on the wing demands a certain 'flying sickle' form. Scientists call this 'convergent evolution' and, in a similar way, the landscapes of the Lake District and North Wales were formed by the same process, at the same time, and have broadly similar histories, so it should be no surprise that they have ended up with scenery and landscapes which support broadly similar habitats, flora and fauna.

3. This may have partly been due to the French Revolution and the Revolutionary Wars, which from 1789 effectively closed off the circuit of the Grand Tour and led to young English aristocrats expanding their horizons closer to home. Continental tourism really didn't recover until after the final defeat of Napoleon in 1815.

kind of inner, spiritual diversion, but little could be further from the truth. In his address to the Geological Society of London in 1830, of which he had just become President, Adam Sedgwick noted that: 'There is an intense and poetic interest in the very uncertainty and boundlessness of our speculations.' Sedgwick had met William Wordsworth in the Lake District in the 1820s and the two men shared a love of long walks, nature and poetry – Sedgwick even contributed an appendix of three letters and later added another on Lake District geology to the poet's famous best-selling *Guide to the Lakes* in 1842, while apologising for the more materialistic concerns of geology at the same time:

> One of your greatest works seems to contain a poetic ban against my brethren of the hammer, and some of them may have well deserved your censures: for every science has its minute philosophers, who neither have the will to soar above the material things around them, nor the power of rising to the contemplation of those laws by which Nature binds into union the different portions of her kingdom.

Sedgwick was a man of the cloth and found some expression of his theology through his geology – for instance he resisted with some vehemence the 'transmutationist' theories of natural selection of his former student Charles Darwin. Meanwhile, Sedgwick's contemporary William Whewell, Master of Trinity College, Cambridge, polymath, another onc-time president of the Geological Society of London, a skilled neologist who first coined the words cathode, anode, physicist, catastrophism and, his crowning glory, scientist, was the author of two books of serious poetry (he was the recipient of Cambridge University's highly prestigious Chancellor's Gold Medal for poetry in 1814) as well as a Professor of Mineralogy. There was more healthy exchange and correspondence between the two worlds of art and science than is perhaps currently given credit for.

Between them, these two worlds conspired to develop the very idea of natural spectacle as being something worth travelling to and looking at. In some ways this was the early beginnings of the tourist trade which, by the late nineteenth century, had resulted in the building of a railway up Yr Wyddfa to handle all the people who were arriving in Llanberis via the branch line from Caernarfon. These tourists would previously have proceeded up the mountain on donkeys. Unlike all the other Welsh narrow-gauge lines that operate today as tourist magnets but were once employed in carrying mineral ore, slate or coal to a port or mainline railhead, the Snowdon Mountain Railway was built solely for the edification of tourists.

The world has come a long way since then, but our fascination with the natural spectacle of our landscape has continued unabated. Nowadays – like the tourist leaflet insisting that Snowdon is an extinct volcano – there is, however, a perception that natural grandeur is not enough, that we have to garnish it a bit to make people sit up and notice.

Thankfully, the audio commentary on the Snowdon Mountain Railway only claims that the mountain is made from volcanic rock, rather than the actual volcano it is sometimes dressed up as. Sadly, though, the rest of the commentary which accompanies you up to the peak really amounts to sixty minutes of typical portentous codswallop disguised as darkness and foreboding. It is an unfortunate symptom of packaged tourist activity, a victim of our allegedly withered attention spans, that the people who run tourist attractions based on the natural wonders of Britain seem to have no faith in their customers' ability to digest magnificence without some kind of superimposed narrative. As the second highest mountain in Wales, the massif of Yr Wyddfa brings with it a heightened expectation of drama and, likewise, a narration on that mountain brings with it a heightened expectation of brooding Welsh poetry.

It is my unfashionable belief that the only proper reaction to Yr Wyddfa should be silence informed by awe and punctuated by an occasional gasp, and anything else is just irrelevant tosh. Wordsworth had it about right in the Mount Snowdon section that draws *The Prelude* to its conclusion, in which he described his climb to see the sun rise from the top of the mountain (though his visit in 1791 was a full century before the advent of the train) when he wrote:

> It was a close, warm, breezeless summer night,
> Wan, dull, and glaring, with a dripping fog
> Low-hung and thick that covered all the sky;
> But, undiscouraged, we began to climb
> The mountain-side. The mist soon girt us round,
> And, after ordinary travellers' talk
> With our conductor, pensively we sank
> Each into commerce with his private thoughts

His party made unexpectedly good progress, however, and instead of seeing the sun rise they witnessed the moon, bathing the mist in its argent light:

The moon hung naked in a firmament
Of azure without cloud, and at my feet
Rested a silent sea of hoary mist.
A hundred hills their dusky backs upheaved
All over this still ocean; and beyond,
Far, far beyond, the solid vapours stretched,
In headlands, tongues, and promontory shapes,
Into the main Atlantic, that appeared
To dwindle, and give up his majesty,
Usurped upon far as the sight could reach.

Poetry aside, you should certainly make the journey by day and what you will get on the slopes of Yr Wyddfa are views – heart-stopping vistas down cwm and corrie and the odd, magnificent revelation when the clouds part. If you're on the train, however, your audio accompaniment will be a cut-price Dylan Thomas spoon-feeding superfluous adornment of something more beautiful than most copywriters can accommodate directly into your ear canal. It is done in the mistaken belief that using a lot of words, placed one on top of another, can make something magnificent even better but you don't make a mountain by piling rock upon rock; a mountain is the edited magnificence of something far too large to understand. The Harlech Dome would have once been of Himalayan height and that shows the Welsh-lite copywriting up for what it is, a bit shallow and silly.

One wonders what Wordsworth, with his geological friendships and interests, would have made of the Ordovician scenery at the time of its creation. A wide sea with volcanic turmoil on the ocean floor and with the occasional cone breaking the surface to erupt. Vents and calderas –

the remains of a cone that has collapsed in on itself following a major eruption – have been found around Snowdonia and the composition of the lavas that have been discovered lead geologists to believe that the eruptions were highly explosive. Another feature of the fiery ring volcanoes was the presence of nuée ardentes, high speed glowing clouds surging down the volcanoes' slopes welding ash and fragments of pumice together. Nuée ardentes can achieve temperatures of 1,000°C and speeds of up to 450 miles per hour (700 km/h).

Wordsworth, no doubt would have been able to put it all into some kind of emotional context. His 1802 preface to the *Lyrical Ballads*, the collection that he and Samuel Taylor Coleridge first published in 1798, contained a definition of good poetry as 'the spontaneous overflow of powerful feelings', which does sound a little volcanic, but the popular perception of his poetry is of pastoral verse, written mostly on the extinct volcanoes and verdant landscapes of the Lake District shales and slates. Ordovician certainly, but post-brimstone also and only once all the fire and fuss has died down or, as Wordsworth himself qualified his 'spontaneous overflow', 'it takes its origin from emotion recollected in tranquillity'.

Throughout the Ordovician, the volcanic sediments are interspersed with more conventional marine sedimentary deposits and conditions were often still ripe for the deposition of shale – the shale that would often later turn into economically important slate both in Wordworth's Lake District and in North Wales. An episode at the outset of the Ordovician was such a period and I learnt once of a particularly interesting area of Wales, a little to the south of Yr Wyddfa, which featured curious fossils that had become caught up in the same movements that had created the Harlech Dome. It was an area of shales, turned to slate but, unusually for slate which will not split along bedding planes, trilobite fossils had apparently survived, after a fashion, presumably rotating a little with the minerals flattened in the Caledonian vice. Even so, I was to expect the trilobites – those of an *Angelina sedgwickii* – to be heavily distorted.

So it was that on a damp and squally warm Welsh day in late summer, I made my way up a steep track that ran along the edge of an ancient wood in search of the twisted fossils. We see the work of one geologist after another in *Angelina sedgwickii* – the first epoch of the Ordovician that contains it was originally part of Adam Sedgwick's Cambrian and Lapworth left it there. In the 1960s it was realised that *Angelina* and her friends had more in common with later fossils and it was moved to Lapworth's Ordovician. The fossil, according to a guidebook I had recently split from a seam of unloved volumes in a second-hand bookshop, was apparently fairly easy to find. It could be done perhaps within a quarter of an hour or so of diligent rock-splitting in an abandoned quarry on the other side of the hill – a statement I accepted completely without question. The book reassured me that I would apparently discover at least one contorted specimen splitting those slates apart. However, in order to do that, I had to find the quarry first and it seemed to be peculiarly elusive for a five-acre hole in a Welsh hillside.

Following the directions, I passed through a gate and struggled up a path that ran between ancient drystone walls which, as is their wont in Wales, were carpeted in moss. The wood was musty, its air thick and oppressive and every warm breath was filled with the cloying scent of chlorophyll before a summer storm. My asthmatic wheeze was accompanied up the hill – as almost everywhere in Wales, or so it seemed – by the nearby whistle and puff of a Welsh steam engine, recently re-purposed from carrying rock to ferrying tourists. It seemed strangely apt, as I was the latter in desperate search of the former.

The path levelled out for a few yards as it crossed the saddle of the hill, then plunged down the other side towards the meadows of a flood plain. On this side of the hill, the geology of the area suddenly asserted itself on the path's character. Where once there were only shale blocks, now the wall incorporated thick slate within its structure and the footpath became strewn with large, loose slates making an unconsolidated scree upon which every footstep was uncertain. I elected to walk down through a claggy streambed that threaded its way from one side of the path to the other, preferring the discomfort of wet feet to a long slide towards certain concussion. This trilobite, I told myself, had better be worth it.

After ten minutes of alternate squelching, slipping and side-stepping my way to the foot of the hill, I found myself in the corner of a field, the expected location of a quarry which simply wasn't there. I checked the map, reread the directions and peered back up into a wild wood of tangled hornbeam and oak. I could just about make out what looked like a face of rock, but it may as well have been on the other side of a barbed-wire thicket. The wood was utterly impenetrable. A mosquito rasped into my ear, just one of hundreds buzzing around like paparazzi on mopeds circling their quarry, while I completely failed to get anywhere near mine. However, a few large boulders of slaty shale had tumbled onto the edge of the meadow and I spent the next twenty minutes with a hammer and chisel breaking rocks, all to no avail. I eventually gave up, downed tools and gave in to a mind-altering mix of perspiration and disappointment. It was only later, with the benefit of 20/20 hindsight, that I found out that I had been looking about a mile further west than I should have been and all my hammering and chipping had been utterly pointless.

I never found the trilobite, so I purchased a good one for just under £3 from a tourist information centre in Shropshire. I decided at the time that geology is all well and fine, but I shouldn't let it ruin my life. On my way back, however, I discovered that, sometimes, just looking at the landscape intent on discovering an exotic animal may lead to surprising results. I caught a brief glimpse of my first hawfinch,[4] so I did at least find one elusive creature hiding among the hornbeam and oak.

4. In Britain, hawfinches are highly elusive birds that usually confine themselves to the forest canopy where they seem to blend in. On the rare occasions you get to see one in plain sight, however, they are quite striking in appearance – possessing an enormous gunmetal beak – and are very easy to identify. In thirty-five years of looking out for one, this three-second glimpse of what one field guide refers to as a 'flying pair of nutcrackers' was my first sighting.

CHAPTER FOURTEEN

A Cambrian Himalaya

Cambrian period – from 488 million to 542 million years ago

Between Harlech and Abermaw, behind the wide coastal plain populated with caravan parks built on drifting dunes and the economic alluvium of the tourist industry, lie one of the hidden treasures of Wales – the Rhinogau mountains.

The Rhinogau are special, but not for any immediately obvious reason. They are not the tallest peaks, nor do they receive anything like the same amount of tourist attention lavished on the better-known mountains of Yr Wyddfa (Snowdon) or Cadair Idris, a dozen or so miles to the north and south respectively, but they are well known and loved among hillwalkers, climbers and those in search of an authentic wilderness. Waist-high heather and an exceptionally rocky terrain conspire to make quick hiking progress in the Rhinogau difficult. A mile in these mountains, runs the walker's maxim, is worth two anywhere else, and this is probably the most important factor in keeping the droves of tourists away although it is aided, to some extent, by the magnetic attraction exerted by one of Britain's highest – and most accessible – mountains only a few miles away. Yr Wyddfa, indeed, was described by the Llanberis Mountain

Outcrops of Cambrian rocks in Wales

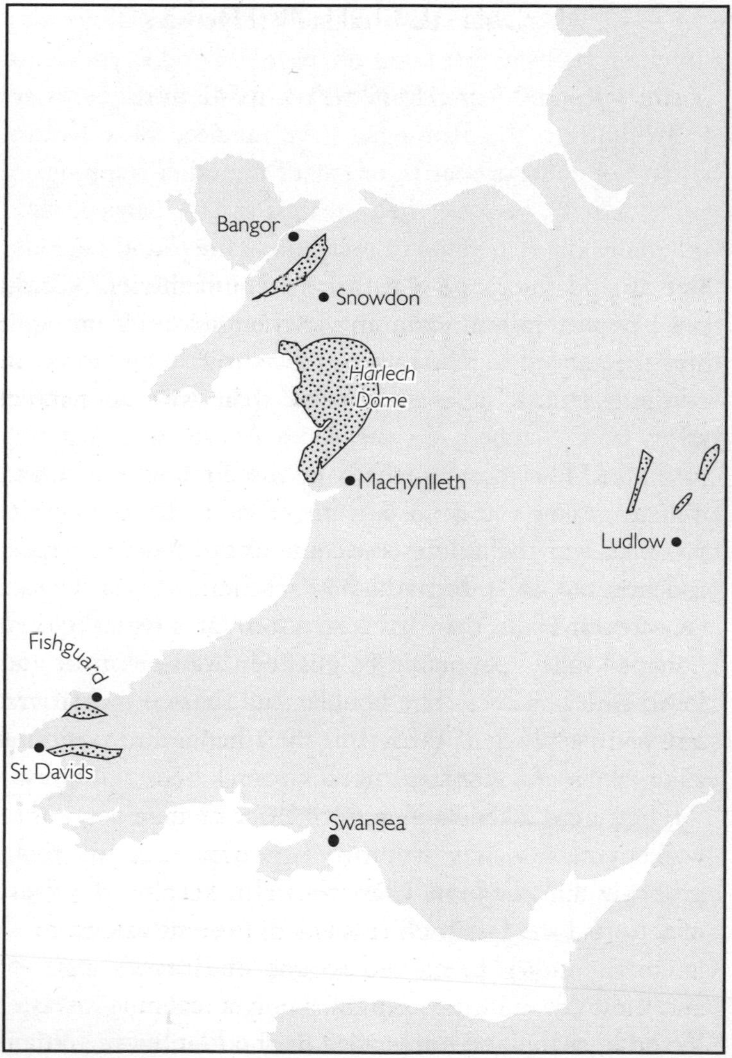

Rescue Team as the 'busiest mountain in Britain' and, to be honest, I've seen quieter shopping centres, an observation which, for me, turned all notions of unfettered wilderness on their head. In fact, it is busy enough even without the ever-so-obliging, but unfortunately expensive, Snowdon Mountain Railway that takes you to the Hafod Eryri Visitor Centre within 60 feet (18 metres) of its summit.

By contrast, the Rhinogau have no need of a visitors' centre, nor context boards, or any of the other trappings of tourist activity, because their greatest asset – particularly if you ignore the rich veins of manganese ore found here and there around the range – is their raw authenticity. Nor do they have any railways running to their summits although there is some aid to the walker in the shape of two tortuous little single-track lanes that wind their way up parallel valleys from Llanbedr on the A496 Abermaw to Harlech coast road. The most northerly of these routes – to Cwm Bychan – takes you on a twisting, 7 mile (11 kilometres) trip up an ever-dwindling lane constrained by no-nonsense, moss-covered walls. In its highest reaches, over cattle grids and past waterfalls, the road creeps through a typical glacial U-shaped valley, populated by gnarled dwarf oak trees and strewn with boulders. The boulders are striated by glaciers, scratched deeply as if carried to their haphazard positions in the talons of a giant, primeval raptor.

If you were asked to pick a wilderness where you might happen across such a cryptozoological wonder, it would almost certainly be here. Unfortunately, you'd be extremely lucky to find the fossilised remains of even the commonest Cambrian oddity here, but around the central core of the Rhinogau lies its slightly younger suburbs of later Cambrian shales and slates with their softer, more verdant landscapes, and in some of these, some curious finds may be made.

The lane follows the tumbling stream of the Afon Artro up 500 feet (152 metres) or so until one of the last of many testing twists and turns reveals a beautiful lake, Llyn Cwm Bychan, as dark as a deep Scottish loch and one of the purported homes of North Wales' own zoological wonder, the torgoch. Not a serpentine monster nor a species of Welsh dragon, the torgoch is, instead, a genetically distinct variety of salmon, the red-bellied char, which has apparently spent so long in the highland llyns of North Wales that it has forgotten how to migrate. The cumulative effect of all of these features – the lake, the legendary, stay-at-home fish, the mossy walls, the gnarled and stunted trees and all the other valley-that-time-forgot stylings – is a sense that you have arrived in real Wales, stripped of the easy-access, coin-operated, mixed blessings of modern life. In a strictly geological sense, you would be correct, because this is one of the locations that helped geologist Professor Adam Sedgwick erect the Cambrian system in 1835.

As you round the last corner on the lane to Cwm Bychan, the road is suddenly perched a dozen metres or so above the northern shore of the llyn on a ledge several hundred metres long, unfenced and narrow enough to worry the most confident of drivers, so you will probably be too busy tightening your grip on the steering wheel to notice the magnificent face of Carreg-y-Saeth on the other side, with its impressive display of brutish bedding planes. Too busy, at least, until you have parked the car in the field at the head of the lake and collected yourself in one of the thoughtfully provided toilets – festival-style Portaloos, only scrupulously clean.[1] From the field, you can safely inspect the rough face of Carreg-y-Saeth that rises from the south side of Llyn Cwm Bychan. It is formed from the classic Rhinogau ingredients, a mixture of greywacke[2] sandstones and shales collectively known, without the faintest trace of glamour, as the Rhinog Grits.

Once you have parked the car, the walk into the Cambrian heartland to discover this gritty upland landscape is relatively easy by Rhinogau standards, but still not exactly a walk in the park. A stone bridge takes you over the Afon Artro which feeds the lake and up a steep path, leading you through the oaks and then lowland moor a further 500 feet (152 metres) up to the Roman Steps, an ancient (though almost certainly not Roman), roughly paved packhorse trail that leads through the Bwlch Tyddiad mountain pass. In high summer, you'll meet a few others on the pilgrimage but a quick visit (allow about one and a half hours to get there and back from the field) to this old track gives you the chance to safely experience the wild landscape of Wales while still being the nearest thing in the area to a short family trek. The packhorse trail goes on to skirt the upper slopes of Rhinog Fawr – at 2,362 feet, the range's third highest peak. It is believed that it once continued further east towards Bala, thence onwards to Chester. The path to the Roman Steps is a more straightforward approach to the spectacular scenery of the Rhinogs than most other routes in the area but,

1. There is a small car park and camping field with Portaloos. There's a nominal charge and an honesty box.

2. Greywacke is a form of sandstone with more than 15 per cent clay or silt in it.

compared to the easiest climb in British mountaineering – from the summit station of the Snowdon Mountain Railway to the top – the Rhinogau is still hard going.

Counter-intuitively, and as we have seen, most of Wales is not, geologically speaking, Cambrian at all. In fact, by far the largest area of the Principality lies under the more recent rocks of the Silurian. At Cwm Bychan, along with the rest of the Rhinogau, however, the later strata have all been removed by erosion, allowing us to look back over 500 million years as if through a very large keyhole. This peek through the ages takes us to the bottom of the Welsh Basin, a depression on the shelf area off the north-west coast of Avalonia where sedimentary rocks accumulated in thicknesses measured in miles. The Welsh Basin, which stretched from the English border to a raised platform running south-west from Anglesey, persisted for hundreds of millions of years and all of the activity in this chapter and most of the last two either occurred in the basin or on the shallower seas at its edges and margins.

The Cambrian system[3] was originally worked out on British soil, just as the two systems that follow it – the Ordovician and Silurian – were also home grown. While the later periods were named after Welsh tribes, Cambria is the Romans' Latinised form of Cymru, the Welsh word for Wales, which derives from an ancient Brythonic word for 'compatriot'. In 1835, after five years of work in the area, Professor Adam Sedgwick erected the eponymous system in a joint exercise with Roderick Impey Murchison, whose various contributions to both geology and controversy span

3. It's worth restating the difference between period and system here: the system refers to the rocks, fossils and other geology of its corresponding period.

the middle part of the nineteenth century like a particularly spiteful marriage – but one punctuated by moments of superb insight and understanding as well as the odd attempted reconciliation.

While Sedgwick worked in the north and west of Wales, Murchison took the south and the eastern border counties. This worked well in terms of the geology as the strata that Murchison investigated and classified were predominately the younger ones of the Silurian period. Sorting them into order, the two men published their joint paper under the title *On the Silurian and Cambrian Systems, Exhibiting the Order in which the Older Sedimentary Strata Succeed each other in England and Wales* in 1835 to the British Association for the Advancement of Science. The paper, which became an early prototype for the entire geological timescale, nevertheless had its weaknesses. The boundary between the two – at the time – contiguous systems was blurry and indistinct and Murchison increasingly made the best use of the confusion, co-opting more and more of Sedgwick's Cambrian rocks into his own Silurian system. Where two systems join usually indicates a marked change in the type of rocks being formed and, by extension, the fossils found within them.[4] Both men were driven to discover the rocks in which the oldest fossils were to be found – signifying the very start of life – and, back then, the upper parts of Sedgwick's Cambrian appeared to contain the first invertebrate fossils. Murchison argued that the Upper Cambrian was really the Lower Silurian and the

4. The type of rock laid down at any one time is itself indicative of the environment it is formed in, which, in turn, has an effect on the type of animal likely to be found there.

friendship of Sedgwick and Murchison suffered a great deal as a result and eventually turned to bitter acrimony.

And it may have seemed like a cruel irony to Sedgwick in 1839, when their friendship was under the most severe strain, that upon the publication of his epic synthesis *The Silurian System*, Murchison chose to write the following dedication: 'To you, my dear Sedgwick, a large portion of whose life has been devoted to the arduous study of the older British rocks, I dedicate this work.' On this evidence alone, it is safe to assume that tact was not Murchison's greatest strength.

It could appear pedantic and childish to argue in such a territorial manner over such matters, but the blurriness of the boundary between the Cambrian and Silurian concealed a deeper problem – namely, that the rocks of the Lower Silurian did not really belong to either the Silurian or the Cambrian system. The question of which rocks belonged where was eventually solved years later, long after Sedgwick and Murchison were safely added to the strata of their respective burial grounds, when schoolteacher and amateur geologist Charles Lapworth erected the Ordovician system in 1879, inserting it like a boxing referee between the Cambrian and Silurian. This wasn't simply a geological fudge to appease two warring parties, Lapworth's Ordovician was a system of real substance with a fossil profile that was distinct from both the Cambrian and Silurian – a fact resisted by the professional survey men until over twenty years later. These three periods, the Cambrian, Ordovician and Silurian, define the structure and landscape of all of mid and north Wales and some of the south-west of the country as well. Their influence is bounded by the edges of the Welsh Basin and extends over the border to the Welsh Marches of Shropshire.

As Sedgwick discovered, North Wales has by far the most complete exposures of Cambrian geology in Britain and with that geology comes the superb scenery of Cwm Bychan and the rest of the Rhinogau – the distinctive landscape that occupies the country between the peaks of Yr Wyddfa and Cadair Idris. In topographical terms, the area comes under the aegis of the Harlech Dome, an umbrella formation of rocks that once stretched high over the Rhinogau. A dome structure of the kind found at Harlech is really just a form of anticline[5] where the compressional forces of mountain building have conspired to raise the beds to a peak rather than a long ridge and where, it follows, the strata all dip away from the centre. When erosion removed the dome's peak, the Cambrian rocks of the Rhinogau were exposed at the centre and on geological maps, this manifests as a rough circle or ellipse of old rock surrounded by concentric rings of progressively younger material.[6] The inward facing Ordovician slopes of Yr Wyddfa and Cadair Idris are merely the amputated anticlinal limbs; the south slope of Yr Wyddfa and the

5. For convenience, it bears repeating here that an anticline is an upfold of strata with the oldest rocks at its core. A simple aid to distinguishing between an anticline and its opposite – the syncline – is that an anticline 'points' up like the capital 'A'. Another, perhaps simpler, way is to think of a syncline as a 'sink' . . .

6. For convenience, it also bears repeating here that the opposite of a dome is a basin – a roughly circular syncline or depression of the sedimentary beds. Its appearance on a geological map is reversed also, with younger rock at the centre surrounded by concentric rings of older outcrops. Counter-intuitively, the rocks of the Harlech Dome were formed in the Welsh Basin.

north of Cadair Idris are both steeper than the slopes that dip away from the centre of the dome.

In these two places, it is evident that erosion has cut across the bedding planes of the dome and formed great escarpments. These have some structural similarity to the chalk cuestas of the North and South Downs which face one another across the now eroded ridge of the Wealden anticline. Similar, that is, except for scale, but even here, the anticlinal structure expresses itself in microcosm around the Rhinogau mountains in the formation of miniature escarpments, one on top of another up the faces of the peaks, giving them their distinctive terraced appearance.

Because the Iapetus was such a wide ocean, the varieties of fossils that are found in Wales, for instance, are completely different from those found a few hundred miles north in Scotland, a state of affairs you wouldn't expect to see had Scotland always been joined to the rest of Britain. Given our modern knowledge of drifting continents, this isn't so surprising now, but it was something of a mystery before the theories of plate tectonics were widely accepted less than fifty years ago.

With the modern Atlantic separating Europe and the Americas – the Old and New Worlds – it is obvious to us that different species of animals live on opposite sides of the pond. Life separates into what biologists call faunal provinces. In the same way that modern Europe and North America are in different faunal provinces, during the Cambrian period, so too were Scotland and the rest of Britain. So, although Laurentia and Avalonia both had trilobites[7]

---

7. Trilobites, I like to think, are the second most famous fossil (the most famous type of fossil is, arguably, the ammonite) that isn't either a dinosaur of

swimming around the seas, the width of the Iapetus at the time of their evolution ensured that they were completely different species. Fossils of the trilobite *Olenellus* found in the Cambrian sediments of Laurentian North West Scotland as well as North America are members of the Pacific fauna, and do not occur in England and Wales.

Meanwhile, for the *Paradoxides* of Avalonia – members of the Atlantic fauna – paradoxically, the reverse is not quite true. One important fact that I have so far omitted to mention about the slow do-si-do of the continents is that Avalonia is named not after the Avalon of Arthurian legend but the Avalon Peninsula of Newfoundland. Just as Laurentia bequeathed Scotland, Avalonia donated a sliver of itself to the eastern seaboard of the northern United States and Canada. *Paradoxides* may also be found – among other places – in the cliffs of the St John River in New Brunswick, Canada. In order to get within a few hundred miles of one another, *Olenellus* and *Paradoxides* were moved together on the backs of continents by plate tectonics – now safely entombed in their respective strata.

As we have already seen, shales – the most common sedimentary rock – are fairly ubiquitous in the British Cambrian, which is just as well for Wales. For shale is essentially petrified clay and the structure of the mineral grains found within it, it could be claimed, are responsible

---

some kind or the missing link between humans and apes. Now wholly extinct as a group that once numbered thousands of species, trilobites were arthropods like modern spiders and crabs. Their closest living relatives are believed to be horseshoe crabs, which they bear a resemblance to and which have not evolved much in form since the Ordovician period, 445 million years ago. There's a representation of a trilobite fossil on the front cover of this book, just above my name.

for the economic fortunes of North Wales, because it will later turn into slate.

The grains of shale are tissue-thin, scale-like laminae forming tiny sheets less than 0.1 mm across and when they accumulate to form shale the sheets lie this way and that on top of one another, like a box of loose Lego bricks all jumbled together. When the shale of Wales was subjected to low-grade metamorphism – that is, metamorphism without a great amount of heat being applied – and was compressed as a result of an episode of mountain-building, the fine laminae of shale line up perpendicular to the force applied and this is what develops slate's die-straight cleavage. The cleavage – the line through which a rock is predisposed to split – is different in slate to the plane upon which the shale was deposited and a roof slate will cut across the bedding planes of the former shale. A typical Victorian slate roof contains hundreds of miniature cross sections of the sea floor over time, with the odd purple or greenish horizontal band revealing the junction of the original beds. You can spend time in large towns and cities looking for prestigious buildings made from imported limestone, examining fluted columns of cathedrals and art galleries for fossils and you may be lucky, but move away from the high-status buildings in the city centre and out to the Victorian suburbs and you'll find plenty of geology nailed to the roofs.

Given free rein and without any unconvenient topographic features to get in the way, cities form like trees, in a roughly circular fashion, each century adding a ring, another circle of development around its girth. Like the geological map of the Harlech Dome, each concentric circle as you travel out from the centre represents a younger formation. In Britain, a great many of our cities and larger towns, when seen from the air, feature a grey ring almost entirely composed of millions of Cambrian roof slates from the quarries of Wales. In this way at least, the economic fortunes of the Principality were originally mud.

Some of the finest roofing slates in the world were laid down as shale in the Cambrian period and then flattened under intense pressure during the Caledonian Orogeny 100 million years later. The products of this pressure are manifest, among other places, in the Llanberis slate, a north-east–south-west belt on the top horn of Wales just to the north of Yr Wyddfa. The quarries at Llanberis are a feat of superhuman proportions. One in particular is a giant's staircase of terraces cut into a hillside, half a kilometre high, that towers above Llyn Padarn and the reservoir of Llyn Peris – another rumoured location, incidentally, for the torgach. The Llanberis slates are usually a greenish-grey, but here and there take on a rather fetching purple sheen and are a particularly fine grade of slate that, in experienced hands, cleaves evenly; the shale that formed it had mineral grains of a very uniform size and the pressure applied to it during the Caledonian Orogeny was apparently 'just-so'.

We find ourselves back at the opening of the Cambrian period, the line beyond which if you were a Victorian geologist, virtually nothing was known. By the close of the nineteenth century, the Cambrian radiation of life had long been noted in the fossil record. It was a particular bugbear of Darwin – who, coincidentally, accompanied Sedgwick as a student during his Cambrian excursions in 1831 a month before he received the invitation to sail on HMS *Beagle* as a naturalist. Darwin would later struggle to explain the sudden explosion of animals and plants around the world in his seminal work on natural selection, *On the Origin of Species*. A tiny hint of the exact nature of it, however, was suddenly brought into sharp focus by one particular discovery on the the other side of the Atlantic in 1909. On the 31 August of that year, Charles Walcott, the Fourth Secretary of the Smithsonian Institute and a palaeontologist with the US Geological Survey, was riding with his family, near the town of Field in the Canadian Rockies looking for beds rich in trilobites along Fossil Ridge, when he uncovered a treasure trove of fossils of extraordinary animals in a layer of black shale. Its colour is important; its very blackness signifies that it was deposited in anoxic conditions – and this absence of oxygen is the principal reason for the excellent state of preservation of the fossils it contains.

The significance of this bed of shale, subsequently called the Burgess Shale, is that it contains the fossils of animals which were completely soft-bodied – creatures that, in the normal course of things, would stand a poor chance of preservation. The Burgess Shale showed the whole world that the Cambrian seas were alive with the most extraordinary creatures imaginable, including one with such a bizarre appearance it was named *Hallucigenia*. Among the weird and wonderful creatures of the Canadian Rockies was a worm a little over an inch long christened *Ottaia*.

Ottaia, however, also had a friend – a mollusc called *Hyolithus*, found in *Ottaia*'s stomach. Now you might find a fossil of *Hyolithus* in Warwickshire if you have a keen eye (it was only 5 millimetres long), but its predator *Ottaia* is totally absent. The question is: is it absent because it wasn't there or because it has not been preserved? Who knows what strange fauna and flora existed in the seas that once covered the British Isles?

Over the years, Walcott excavated 65,000 fossils from the Burgess Shales and spent the rest of his life putting them in order. The quality of preservation alone qualifies the fossil beds as a Lagerstätte,[8] a distinction it shares with a number of other Cambrian shales. This presents us with a paradox – namely that the fauna and flora of the Cambrian is much better understood than many more recent periods.

Cambrian geology is relatively well known in Britain and although it is highly unlikely that we will ever find find any Lagerstätten from the period in this country (especially in Wales, which is relatively poor in terms of fossils of this age, certainly in comparison to England), unusual finds are, nevertheless, still made. Just east of the A49 that runs from Church Stretton to Shrewsbury in Shropshire lies one of Britain's most important geological and palaeontological sites, Comley Quarry. In 1888, Charles Lapworth once again settled a geological arms race in which Murchison was entangled, that of finding the earliest fossiliferous strata. He did this by describing the oldest Cambrian fossil, a trilobite, *Callavia callavia*.[9] It was found in a rare exposed sequence of Lower and Middle Cambrian sandstone and limestone in the quarry. It is thought to be approximately 550 million years old, right on the very edge of the Cambrian: almost Pre-Cambrian, in fact.

---

8. Just in case you've forgotten – a German term for a sedimentary bed rich in fossils. Its literal German translation is 'place of storage'.

9. A specimen of this trilobite – originally known as *Olenellus callavei* – was originally collected by Charles Callaway.

CHAPTER FIFTEEN

Of Avalon and Avalonia – the Terranes of England and Wales

Pre-Cambrian eon – from 542 million to 1 billion years ago

On a gusty day in April, the southern corner of Ynys Môn – pronounced 'Uh-niss Morn' or, in English, 'Anglesey' – looks benign enough. Indeed, its gently undulating plateau of green fields, punctuated by stone walls and thorny hedges, cannot help appear anything but mild in comparison to the neighbouring landscapes of which it has such dramatic views, Snowdonia and the Pen Llyn (Lleyn Peninsula). Anglesey is neither the Wales of lofty peaks nor one of steep parallel valleys alive with the hum and clatter of heavy industry; indeed, this is a landscape that looks more like the placid scenery of mid-Cornwall than the drama of everything that Anglesey has to the south and east.

Like the gneiss of the North West Highlands, this uniform, low plateau belies the qualities of the rock which forms it; as in the North West Highlands, large areas of the flattest parts of Anglesey are made of a complicated assortment of ancient igneous and metamorphic rocks that are very hard indeed. Yet its low relief is a reflection not of

Places of geolgical interest on Ynys Môn (Anglesey)

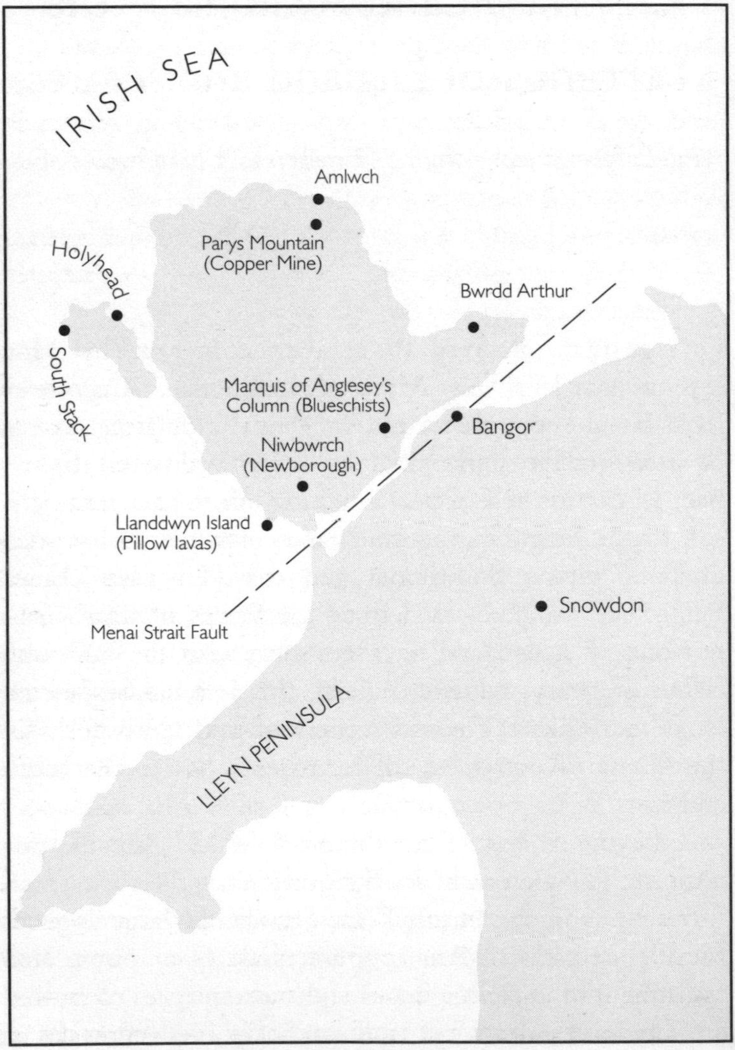

the rocks themselves, but of how they were eroded. In this case, a long transgression of the sea has made much of Anglesey an ancient wave-cut platform.

Given the remote, romantic nature of islands – in history, a stretch of rough water was often as good as a couple of hundred miles of trackway to impede communications – it is not surprising that an island on the Celtic fringe of Wales and, therefore, at least three levels of abstraction from mainland Britain, is considered to be such a place of legend. Such remoteness, forced upon Anglesey by distance, a wall of mountains and, finally, a short but treacherous sea crossing, can only foster uniqueness – perhaps more so when you consider its epithet was the 'Island of the Druids'.

Like many places in Wales and the west, it also claims associations with King Arthur, the most extreme and questionable of which is that this is his last resting place, Avalon. It is a legend it shares with nearby Bardsey Island (literally the 'Island of the Bards', i.e. druids) off the western tip of the Lleyn Peninsula, though Bardsey wins by a country mile in the comparison. There is a large hill fort over 530 feet (161.5 metres) high called Bwrdd Arthur, or Arthur's Table – north of Llanddona on the eastern spur of Anglesey – which features a table-and-chairs style formation of a limestone pavement. The views are terrific and its defensive qualities are obvious, but the 'furniture' is no more Arthurian than any of the other natural 'round table' formations scattered around Wales and the west. Indeed, the Arthur–Anglesey association is so slight and flimsy, it is even more nebulous and insubstantial than most Arthurian legends, which themselves often appear to have been woven from nothing more tangible than the distant memory of mist.

There are scattered and inconclusive references to

Arthur before Geoffrey of Monmouth, the twelfth-century clergyman, wrote his far-fetched chronicles *Historia Regum Britanniae* ('The History of the Kings of Britain') and *Vita Merlini* ('The Life of Merlin') and created the familiar literary persona of Arthur. According to Geoffrey, Avalon, which comes from the Welsh word for apple – *afal* – was known first as Insule Avallonis in *Historia* . . . then Insule Ponorum, 'the island of the apples' in *Vita Merlini*. Anglesey – runs the ever-tenuous thread – was, in Roman times at least, famed for its production of apples.

Mythical kings aside, Anglesey did represent a stronghold of Welsh freedom and real druidry – although, like Arthur, it is difficult to be precise about the term, littered as it is with the detritus of the Romantic revival that inspired the daft notions of medieval courtly love and chivalry. The Romans, however, certainly encountered resistance from Anglesey – an island on the other side of a mountain range made life difficult for the legions which sought to conquer her. The Roman historian Tacitus gives an account of a particularly vicious battle at the Menai Straits between the Celts and Gaius Suetonius Paullinus:

> By the shore stood an opposing battle-line, thick with men and weapons, women running between them, like the Furies in their funereal clothes, their hair flowing, carrying torches; and Druids among them, pouring out frightful curses with their hands raised high to the heavens, our soldiers being so scared by the unfamiliar sight that their limbs were paralysed, and they stood motionless and exposed to be wounded.

Paullinus led his troops to a military and spiritual routing of the Druids and he took it upon himself to destroy the groves that were so sacred to them. Although it was initially successful, the 60 AD invasion was not consolidated until the Roman Governor Agricola inflicted a punishing and conclusive triumph over the Celtic tribes in 78 AD, and the island was taken for good. On the Strait side of the island, the memory of those campaigns are forever preserved in the names of fields; close to Brynsiencyn, two howl their names as 'The Field of the Long Battle' and the 'Field of Bitter Lamentation'.

In the verdant fields of southern Anglesey today, these are the only hints that violence ever darkened the island. The memory of war is as intangible as the shadows of clouds that race across the fields, lending the low, rolling terrain a dynamic and animated air. So it is that, in terms of landscape as well as history, Anglesey today may be a much more settled prospect than the rest of Wales, but Avalon it is not. There is a connection, however – and a less tenuous one than those woven by the legends that seem to hang heavy over all genuine Celtic landscapes. The connection is not so much with Avalon as Avalonia – the ancient micro-continent on which all of England and Wales were created.

The violence of Anglesey's history was not just that inflicted by invading armies – in fact, in common with the rest of England and Wales, its very birth was a pyrotechnic extravaganza of the first order, although the exact nature of the extravaganza is a scientific mystery. For a small island, Anglesey is very complex and, like the North West Highlands, would merit an entire book of explanation in itself. But that all depends on whether that explanation was actually forthcoming, because there is no real sign of it yet. Unlike the North West Highlands – the subject of decades of controversy and political wrangling – the academic work concerning Anglesey has been more civilised and it is not my place to suggest that that might be a reason why the exact geology of Anglesey hasn't really been settled.

What is known is that the island boasts the largest exposure of both metamorphic and Pre-Cambrian geology in southern Britain and is structurally formed of three basic units: a strip of granite and gneiss forms the first, a rare accumulation of a unusual metamorphic rock called blueschist the second and a collection of sedimentary, volcanic and metamorphic rocks make up the largest and third of the groups. All of these groups are, in places, overlain by later rocks such as the limestone that makes Arthur's dining arrangements at Bwrdd Arthur.

The exact relationships between the three units is not really that clear. In fact, the connections are so unclear that they are the subject of academic debate too involved and too drawn out to go into here. Like the North West Highlands, the problems are tectonic and a straightforward synthesis of the island's ancient history is made difficult by a complex network of faults.

In summary, though, there's a general trend on Anglesey that rocks become steadily more altered by metamorphosis the nearer they are to the mainland, reaching a peak in a belt of the second group – the blueschist. Despite its name, it is usually dark grey (though you may detect, as I did, a deep purply blue sheen) but it is not for its appearance that it is celebrated. At 560 million years old, the blueschist of Anglesey may be the oldest so far discovered in the world. But even that boast hides the important fact that it was formed under very specific conditions; conditions which, outside the laboratory, are only usually found where an ocean plate is being subducted. The ancient blueschists may be rarities in an international context, but on Anglesey they are very easy to find, close to the tourist trail at Llanfairpwllgwyngyll-gogerychwyrn-drobwll-llantysiliogogogoch (the locals call it Llanfair PG or even just Llanfair to save time).[1]

On the south-eastern edge of the town, overlooking the A5 and the Holyhead to Bangor railway line, stands the Marquess of Anglesey's Column, a 88.5 feet (27 metre) high memorial to a Waterloo hero built from local limestone upon an ancient hill fort. The hill fort rests on the blueschist and there are a number of outcrops in the woods below the column. If you visit on a clear day, you may as well take advantage of the breathtaking views of Snowdonia, the Menai Strait, its bridges and most of Anglesey by climbing up the 115 steps inside the column.[2]

The third major unit of geology on Anglesey is a series of rocks which are all lumped together as the Monian Supergroup (confusingly, this has nothing to do with the Moinian in North West Scotland – Mona is the Latin name for Anglesey). In the north of the island, the Monian includes possibly one of the most astounding shows of rock-bending acrobatics you are ever likely to see.

South Stack is an island that lies 30 metres off Holy

---

1. Llanfair PG is the town that officially renamed itself in the late nineteenth century to achieve fame as having the railway station with the longest name in Britain. (The name means *St Mary's Church in the hollow of the white hazel near to the rapid whirlpool and the church of St Tysilio of the red cave.*)

2. The column is open all year, is well signposted and there is a small admission fee.

Island, itself only joined to the 'mainland' of Anglesey by road and rail bridges – and the strata there include sandstones called turbidites. Originally laid down by furious ocean-floor currents, they now twist this way and that, in contortions and buckles as if reliving their turbulent origins. Standing on Holy Island looking across to South Stack, it can feel as if you are standing at the centre of a lithified hurricane. Even *The Pre-Cambrian Rocks of England and Wales*, the weighty scientific tome published by the Geological Conservation Review of the Joint Nature Conservation Committee, gets dangerously close to 'gushy' about South Stack: 'The cliffs around South Stack Lighthouse display some of the most magnificent exposures of folded metasedimentary rocks in Britain.'[3]

The feeling is compounded, particularly in summer, by an awe-inspiring colony of seabirds including guillemots, razorbills, fulmars and kittiwakes, wheeling and diving between sea and sky. The twists and turns of the turbidites that are interbedded with mudstones on the cliffs of South Stack provide ample nesting opportunities not only for these seabirds – which only come to land to breed – but for the less pelagic gulls, as well as ravens and Britain's rarest member of the crow family, the chough.[4]

Such dramatic landscapes as those of South Stack and the surrounding coastline are rare on Anglesey and almost all of them are to be found on the coast. The island's run-of-the-mill interior might even feel like a bit of a let-down

3. *Pre-Cambrian Rocks of England and Wales*, Geological Conservation Review Series 20, by J.N. Carney, J.M. Horak, T.C. Pharaoh, W. Gibbons, D. Wilson, W.J. Barclay, R.E. Bevins, J.C.W. Cope and T.D. Ford.

4. South Stack is also the location of an excellent RSPB nature reserve.

if you've arrived there via the high drama of Snowdonia or the equally stirring scenery of the seaboard of North Wales but, like an indifferent picture set in a beautiful frame, the coastline more than makes up for it. It is not for nothing that almost the entire 125-mile (200-kilometre) coastline of Anglesey is designated as an Area of Outstanding Natural Beauty. However, it isn't only aesthetics that reach their full potential at the coast because, just like South Stack, it is here that the structure of the island is revealed as well.

I set off for a Forestry Commission plantation near the southern tip of the island, where a number of ancient secrets would be revealed. I was heading for a spot a few miles to the south-west of the village of Niwbwrch (Newborough) on the southern corner of the island. On my way there, up the minor road from the centre of the village, my eye was drawn to St Peter's Church, a building the size of a chapel. It was apparently built in the early fourteenth century, something you wouldn't guess by looking at its pebble-dashed exterior, and is set in a square plot next door to the site of a twelfth-century royal palace at the head of the Gallt Bedr ridge forming the watershed between the rivers Cefni and Braint.

Like the North West Highlands, the landscape of this part of Anglesey has a definite 'grain' to it and this ridge follows the bearing of that south-west–north-east 'Caledonian' grain for good Caledonian reasons, being part of the collision between Avalonia and Laurentia around 400 million years ago. This grain is even evident in the very shape of Anglesey, the southern portion of which has a coastline that resembles the bottom half of a diamond, from the corner of which one coast runs north-west towards Holy Island and the other north-east up the Menai Strait. This orientation of physical features has had its effect in that part of the island closest to the rest of north-west Wales, where roads and rivers run either roughly parallel or perpendicular to the line carved by the Menai Strait. The Strait marks the approximate course of a very deep fault line separating Anglesey from the mainland and the fault is one of a complex that run through the island in the same direction. Faults are lines of weakness – the tectonic forces that cause them crush the rocks in a fissure, which can then be exploited by glaciers, rivers and the elements. The Menai Strait was formed in this way. At many beaches and a few low cliffs along the Strait you can see the sands, gravels and shingles dumped there by glaciers – and until sea levels rose around 10,000 years ago the Strait was two river estuaries, one running south-west, the other north-east, with a watershed between them.

It is easy to see how a river or glacier would exploit the weakness of a fault in carving a valley for itself – but it is less clear how the undramatic topography of this part of the island in modern times would confront human life with any kind of serious challenge. It is almost as if the roads were laid out not in spite of the landscape – as happens in truly hilly terrains – but in reverence to it. In those terms, the geology is faintly remembered in the landscape and the pattern of human settlement has picked up the underlying reality rather than the actual geography. It is an excellent example of how man has followed the weakest of lines imposed upon him by the landscape.

Because of its position, some way from the village centre, St Peter's looks curiously divorced from Newborough, but the church is by no means estranged and, indeed, shows, through its well-stocked graveyard, that it has been in continuous use for the last 700 years. What particularly took my eye, especially in the light of the land's subtle influence over all who live on it, was the matter of its influence on those who chose burial there. The headstones of its most recent occupants carry information I have never seen before on a memorial or tomb. Inscribed on almost every recent gravestone was the former address of its resident, along with the usual assemblage of names, ages and a comforting passage of scripture. As they lay there 'asleep' – as the euphemism has it – they are forever connected to their lives; not just as memories for families and friends but also tangibly, for anyone who cares to look, to a special location in the world they left behind, their home. It may well sound macabre and will probably, at some point, become illegal under data protection laws, but I've never seen this rather touching connection between people and place demonstrated quite so well as in that quiet cemetery. I still wonder whether it is something unique to Newborough, Anglesey or Wales, but perhaps it is Wales: of all the native languages of these islands, Welsh is unique in having its own word for that special kind of yearning, that homesick longing to be back home. In Welsh it is called '*hiraeth*'.

After a walk of a couple of miles through Newborough Forest, I arrived at Llanddwyn beach and could immediately see why people would never want to leave either Newborough or Anglesey. The warren of dunes, the forest and the long sandy beach are all absolutely breathtaking; fantastic for bird watching, walking and cycling, the forest has a colony of rare red squirrels – Anglesey is the only Welsh location where there are no grey squirrels – and the beach is great for kite-surfing. It has a large car park with toilets and is popular in summer, but the beach is so big there is probably always space available.

In April, however, it was deserted, save for a single kite-surfer ploughing through the waves to the south. There is nothing quite so evocative as a walk on a long blustery beach and, at Llanddwyn, the white gold of the sands is set against the stunning backdrop of the dark-grey mountains of the Pen Llyn and Snowdonia, while the fluting of curlews, the piping of oystercatchers and the rough ratchet call of ravens wheeling overhead add to the barrage of sensations that connect us to the natural world. You don't have to be Emily Brontë to recognise that there is something about a windswept and lonely landscape that practically screams forlorn souls and unrequited love, so it is no wonder that Ynys Llanddwyn, literally the 'island of the church of St Dwynwen', the Welsh patron saint of lovers, is such a romantic spot.[5]

Drinking in the light, heady atmosphere of romance is all well and good, but I was on the lookout for a much heavier kind of object altogether, because a particular kind of landscape occurs in abundance around this romantic island and it hints at a far from straightforward sequence of events in the history of Anglesey.

Like the serried stands of Newborough Forest's Corsican pine, from a distance the pillow lavas of Llanddwyn appear black against the sand. But on close inspection they have something of the forest about them – a deep glaucous

5. The Welsh equivalent of St Valentine, St Dwynwen's Day is on the 25th of January and Welsh speakers choose her day over that of her Roman counterpart to send cards and flowers to the object of their affections.

green shade and a lustre almost like the bloom of a grape. The whole outcrop is suggestive, in colour and texture at least, of weathered lead on a church roof.

Pillow lavas are the result of volcanic eruptions under the sea where a tongue of hot magma oozes out from a fissure – or vent – onto the sea floor. The lava forms a skin in the cold water almost immediately, while it continues to flow in from the vent, inflating the skin further until it forms a balloon-like lobe. When the internal pressure becomes too great, the skin is broken and another tongue of lava is started, leaving a collection of squashed and crowded pillows behind. Each one here is the approximate size of an inflated party balloon, but they can be up to a metre high. The details of their formation are so well known and understood because their modern equivalents have been filmed at the moment of creation in the mid-Atlantic and the circumstances under which the Newborough pillow lavas were formed seem to be exactly the same. It is believed that they mark where the Iapetus Ocean opened up nearly 600 million years ago. Today, the extruded pillows sit on the beach in the same way that colossal piles of petrified horse manure would if they were discharged from an animal the size of a rugby stadium.

The deep green of the lavas is set off in some of the gaps between the pillows by a glorious rusty red chert. Chert, an opaque quartz, is a fine-grained rock rich in silica and closely related to flint – the only distinction being that flint is dark and is found as nodules in chalk and limestone. Chert is chemically the same (except for any impurities that might lend it some colour) but fractures in a different way to flint. It not only occupies the space between the pillows but also some of the vacuoles left behind by trapped gas within the lava. The chert at Ynys Llanddwyn which, because of its chemical composition is called 'jaspery chert', was formed when silica and iron were precipitated out from hydrothermal fluids[6] associated with such a volcanically active ocean bed.

Because of their unique shape – rounded at the top and tapering to a point at the bottom – pillow lavas perform a very useful service in geology. In a refreshingly and unusually simple (for geologists, at least) phrase, they are referred to as 'way-up' markers. This comes in handy in complicated areas where thrust faults and extensive folding have occurred and getting a grip on the stratigraphy is difficult; a way-up marker will at least tell you whether the rocks you are looking at are the right way up. And if you're looking for complex geology, you couldn't make a better start than Anglesey. Some geologists have spent decades trying to work it out and there is still plenty of work to be done.

Geologically speaking, the island is peculiar not only in the familiar adjectival meanings of the word, baffling and perplexing, distinctive and unique, but is also believed peculiar in a modified form of its original sense – that of a noun that describes a parish outside the jurisdiction of the diocese in which it lies. On Anglesey, that peculiarity is thought by some scientists to be geological rather than ecclesiastical; Anglesey, according to the argument, is at the eastern end of a tract of crust with a geological history that is part of, but might be distinct from, the rest of Wales – almost as if the island has been adopted.

---

6. Hydrothermal fluid is mineral-rich water heated by volcanic activity in the earth's crust.

In the 1970s, geologists noted that the rock successions of the island have little in common with the rest of Wales over the other side of the Menai Strait and some of them believe that that is because it was formed on a different terrane – the Rosslare–Monian Terrane. Terrane is a specialised term for a distinct block of crust, delineated by faults, with its own history. The idea corresponded fairly well with the situation as it was seen; this terrane, according to the theory, was joined or 'sutured' at some point to Avalonia, the micro-continent on which the rest of Wales and all of England are formed. In the parlance of geology, the Rosslare–Monian Terrane was said to be 'suspect' in relation to the Avalonian Terrane. If Anglesey is part of a suspect terrane, presumably it was formed at some distance from Avalonia and only joined itself to Wales at a later point in its history. Current theory suggests that Anglesey only came into such close proximity to Wales around 400 million years ago and that join today is the major tectonic boundary of the Menai Strait Fault System.

Yet, there is an emerging viewpoint that this exciting idea unfortunately stems from the wrong interpretation of the evidence. The idea that Anglesey was at one time adrift on its own in the Iapetus Ocean, say the detractors, was based on good science at the time but a lot more has been discovered since, particularly with regard to the ages of the rocks on the island, that conclusively rules out the theory. In the currently held view, the two terranes were formed in relative proximity to one another and moved together laterally rather than as a head-on collision.

Edward Greenly would have probably loved the idea of a Monian micro-continent. He spent twenty-five years surveying the geology of Anglesey at an awe-inspiring 6 inches to the mile – the one inch to the mile maps that resulted in 1919 remain essentially unaltered to this day. He was one of the survey men appointed by Archibald Geikie to assist Peach and Horne in unravelling the North West Scotland Zone of Complication that eventually became known as the Moine Thrust Belt. He was so taken with the ideas behind thrust tectonics he found it wherever he went in Anglesey – even if much simpler explanations were appropriate.[7]

One of the features contributing to the idea of a drifting Anglesey was the compelling evidence on the island for a deep trench of the type found where ocean crust converges with continental crust – a subduction zone. At the foot of the Marquess of Anglesey's Column at Llanfair PG, the abundancy of otherwise rare blueschists add further grist to the mill, but there are other tantalising remnants that point to subduction and Greenly was the first to document one of them. When one plate is forced under another, it is almost always the plate that the sea floor rests upon that is subducted under the continental plate, and rocks lying on the sea floor along with shallow water sediments tumble in a cataclysmic underwater landslide into the trench's abyss. There is an excellent example of such a landslide in Anglesey

7. Like the North West Highlands, because of its extraordinary geology, Anglesey has had an almost magnetic attraction through history for scientists, students and the odd contemplating cleric or two. Without doubt, of all the people to wander along its shores, railway cuttings and quarries, Edward Greenly contributed the most. Six years surveying in the field with Peach and Horne in the North West Highlands were excellent preparation for Anglesey's complex successions of metamorphic outcrops, while he seems to have always held on to the enthusiasm he had after a mountain-top epiphany when climbing Cadair Idris with friends and leaving a career in law behind.

and the scale of this submarine pile of rubble – or, to give it its scientific name, olistostrome – is astonishing. It is believed to have covered thousands of square miles and included all sorts of different material of wildly different sizes jumbled together. Greenly gave it the rather grand name of the Gwna Mélange but was at a loss to sum it up adequately except to say that it was 'quite indescribable'.

On the Lleyn Peninsula, to the south-west of Caernarfon, there are patchwork cliffs formed from the Gwna Mélange which include boulders the size of a palacc whilc some 'fragments' in the Mélange are estimated to be a kilometre or more in size. Back on Anglesey, one of the places where a fraction of this haphazard heap is exposed is Llanddwyn Island. At the far south-western end of the island at Porth Twr Bach, large red, pink, green and white boulders appear in an assemblage that looks like nothing so much as a feature of a Chelsea show garden. It is both utterly unexpected for the casual visitor and a highlight of any short ramble over Llanddwyn Island.

The Pre-Cambrian outcrops of Anglesey and Scotland are not the only places in Britain where Pre-Cambrian rocks occur, however. Though nowhere near as old as the basement Lewisian gneiss, isolated blocks of ancient sediments occur also in a few parts of the English Midlands. As we have already seen, Roger Mason discovered his *Charnia* fossil in the Charnwood Forest near Leicester in ashy sediments about 560 million years old, but further west there is a collection of isolated Pre-Cambrian outcrops in Shropshire, Hereford and Worcestershire. In Shropshire, in particular, a line can be drawn on the ground and all to the west is Cambrian, Ordovician and Silurian Wales, while to its east lie the generally younger English

Midlands. Where these two meet, a Pre-Cambrian inlier – an island of ancient rock surrounded by more recent deposits – is exposed. On a map, it is quite easy to spot part of the line that marks the eastern boundary of this inlier; it even shows itself up well on the pages of a road atlas in the shape of the A49 south from Shrewsbury. On a relief map it passes down a long valley – the Stretton Gap – between the whaleback ridge of Caer Caradoc and the long hills of the Wenlock Edge on the east and the area of Pre-Cambrian high relief to its west called the Longmynd; and on a geological map we find that all of those features follow another die-straight feature called the Church Stretton Fault.[8]

The line of the Church Stretton Fault is not the only fault that shows itself so spectacularly in the landscape. The Great Glen in the Highlands is another, the line of the Menai Strait and the orientation of streams, coast and human geography are, as we have seen, readily apparent in Anglesey. In North Wales, just as in its close neighbour, Anglesey, you can see a distinct south-west–north-east Caledonian trend clearly in north, west and central Wales and, once again, it is plain enough on something as basic as a road atlas. The A494, for instance, between the

8. Here, as elsewhere in Shropshire and the Welsh Marches, those with an enthusiasm for the landscape are well catered for and encouraged in their interest by plenty of leaflets, pamphlets, visitor and tourist information centres manned by well-trained staff with more than a passing familiarity with the natural environment. Among these is the Shropshire Hills Discovery Centre in the village of Craven Arms on the A49, an excellent multi-use building housing a library, exhibition, gallery, cafe, bookshop and tourist information centre that also features a free car park, a picnic area, twenty acres of meadows, a riverside walk and a number of paid activities to keep the kids interested.

village of Druid and Dolgellau marks the trend well as it carves a 27 mile (43 kilometre) long primary route incision down past the small town of Bala and its die-straight, 4 mile (6.5 kilometre) long eponymous lake. Llyn Tegid ('Lake of Serenity' is an approximate translation of its Welsh name) is supposedly the haunt of a mythical monster 'Teggie', while its waters are actually occupied by pike and perch, as well as yet another Welsh piscatorial wonder, the gwyniad.

The gwyniad is a rare form of whitefish – a landlocked herring – only found in Llyn Tegid. The reasons for its capture are completely geological – the same reasons, we are led to believe, that Teggie finds itself there; for, like its mythical counterpart, the gwyniad was left stranded in the lake at the end of the last glacial period of the most recent ice age, 10,000 years ago. The gwyniad is increasingly under threat and is being squeezed between the depths and shallows of its only home in Llyn Tegid. Algae growing as a consequence of fertiliser run-off is lowering the oxygen levels in the depths of the lake and warming of the shallower waters in which they lay their eggs in summer is confining them to an increasingly limited section of the water column. An aggressive interloper, the ruffe, introduced to Llyn Tegid in the 1980s, which eats the gwyniad's eggs and young, is not helping matters. In 2003, a project was started to seed more suitable lakes in the area with young gwyniad, as its days in Llyn Tegid look numbered.

When I started this account of Britain's landscape, I never dreamt that I would reach a point where I compared the human condition with that of a landlocked Welsh herring but here we all are, like the gwyniad, victims of geology. We should be grateful as well, for our world is driven by the resources at our disposal and all of them, with the singular exceptions of the Sun and the odd wayward meteorite, come from the ground we stand upon. Over the course of this book the geological record has bloomed along with the chances of catching chunks of our landscape during its various phases of formation, so it is perhaps a good moment to stop and consider just what we have found as we moved up and down the eons, eras and epochs, back and forth through time.

While we might not get much of a look-in on the world stage of calamities in modern history, we have seen that the landscape of Britain wasn't always this quiet. Indeed, this Britain has been one of extremes. There have been volcanoes and massive earthquakes, colliding continents, lush, tropical jungles, deserts and warm shallow seas. There are fault lines and mile-high glaciers, explosive volcanoes and mountains of a Himalayan bent. All, ultimately, upon the same ground we stand on today.

In the light of that history, maybe I shouldn't have been surprised that I was so overwhelmed by the British landscape on that car journey from Cornwall to Wiltshire. Since that journey, I have gone on many more to look at this feature and that, from Lochinver to Land's End and, if anything, I feel more strongly about the beauty, grandeur and sublime charm and majesty of the landscape. It is, after all, very dramatic, a perfect expression of the forces that formed it.

Between the planetary convulsions, the eruptions, the rifting and compression, the pyrotechnics and geological violence that seem to book-end Avalonia's long voyage from the south to the north and Scotland's astonishing journey through time, something very important was happening: the birth of a remarkably diverse and interesting landscape. From flat lands and wide horizons to sublime views and extraordinary vistas, Britain has it all.

POSTSCRIPT

A Quick History of the World Before Britain

Pre-Cambrian – from 4.6 billion to 3 billion years ago

Accreting from a disc of dust orbiting a new main-sequence,[1] yellow dwarf star approximately 4.54 billion years ago, our Earth is almost a third of the age of the universe. That disc – the protoplanetary disc – was the remains of what was left over when one small part of a giant molecular cloud several light-years across was kick-started into contraction by the shock wave from a nearby exploding star. This shock wave moved the molecules closer together, creating a pocket of over density in the cloud.

Forced out of equilibrium, the over density started to contract under its own gravity which, in turn, started a chain reaction attracting more mass which further collapsed and so on. As the cloud contracted it spun faster, much as an ice skater can increase the speed of a spin by bringing their arms and legs closer to their body. This contraction brought the diffuse atoms closer together, where they

1. For an explanation of this term, bear with me and read on.

started to collide much more frequently. Colliding atoms produce heat and over a relatively short time – about 100,000 years – the centre of the disc became hot enough to turn into an early kind of star called a T-Tauri: a star entirely powered by converting the gravitational energy of matter falling into it into heat.

After about 50 million years of further contraction and gravitational heating, temperatures and pressure gradually climbed high enough for nuclear fusion to occur. In this context, nuclear fusion is the process where the nuclei of hydrogen atoms are brought so close together and at such high temperatures that they overcome their general tendency to repel one another at a distance in favour of a much stronger force of attraction that only works in very close proximity. Hydrogen nuclei are then bonded together to form helium and release vast quantities of energy – energy then radiated and convected out from the core through the interior layers of the Sun to form heat and light. The fusion also overcame further gravitational contraction of the Sun and it began what is known as its main sequence. Around the young Sun lay the remains of its formation, the solar nebula – the spinning protoplanetary disc from which planets began to gather.

The first moments of a planet formed in a swirling cloud of dust, rocks, asteroids and planetesimals, is as precarious as it sounds and, indeed, the Earth was perhaps only as young as 23 million years old – a mere moment in geological time – when it is believed that it was hit by a planet the size of Mars. That hypothetical planet – named Theia by scientists – came off worse in the impact. Smashed into countless pieces, it lost most of its iron core to the Earth but then formed an orbiting ring of dust and rock which gradually reassembled what remained of itself from millions of fragments and began to orbit the Earth as the Moon.

The earliest period of time, from the Earth's formation to a point variously fixed between 3.8 billion and 4 billion years ago is known – by some scientists, informally at least – as the Hadean eon. As we have already seen, geologists use a cascading series of time spans to mark the ages of rocks, geological events and fossils over the history of the Earth. They arrive at the ages by using various techniques to divide deep time into four eons, which are further divided into eras, then periods, then epochs. We are, for example, living in the Holocene epoch of the Quaternary period in the Cainozoic era of the Phanerozoic eon. The Holocene epoch goes back just 10,000 years, while the Quaternary period extends 2 million years into the past. The Cainozoic era started 65 million years ago – when the last of the dinosaurs were wiped out – and the Phancrozoic is the shortest of the four eons at just 542 million years old, representing only the last 12 per cent of the planet's history.

Phanerozoic translates from its Greek roots as 'evident' or 'visible life' – an acknowledgement that our own eon's rocks contain abundant evidence of life in the shape of fossils of large organisms. Sandwiched between the Hadean and the Phanerozoic lie two eons: the Archean, which starts from the end of the Hadean; and the Proterozoic ('former life') – the eon that followed the Archean and was immediately before our own. The first three eons used to be regarded as just one, despite covering about 88 per cent of the history of the planet, an eon which was variously known as the Pre-Cambrian or, occasionally, the Cryptozoic. As more discoveries were made and more information emerged – particularly in the Proterozoic and Archean eons – scientists subdivided this huge slab of time into more manageable chunks, but the figures still speak volumes of just how much is known about the Earth in the deeper reaches of time. The Archean is at least 1.8 billion years long, the Proterozoic nearly 2 billion, while, as has already been noted, the Phanerozoic – our most recent eon is not much more than 500 million years in length.

At the other end of time, not all geologists even recognise the Hadean eon, not least because the line between it and the rest of the Earth's history is something of a moving target. The boundary at the lower end of the Archean indicates, as it was meant to, a point in time before which there is no geological record so, as more is discovered, the further the line goes back. The Hadean is, therefore, regarded stratigraphically as something of a void, and its delineation defined purely by an absence of knowledge.

Whereas every other eon of the Earth's history is split to a greater or lesser extent into further units based on specific formations of rocks or the presence of certain kinds of fossils, the Hadean is so distant and mysterious that those kinds of subdivisions are impossible – not because very little happened, but because the evidence has now been obliterated. Like the autobiography of a clean-shaven, wholesome celebrity which skips over a wild and misspent – yet interesting – youth, it is almost as if the planet has edited its own history.

All of these difficulties make the Hadean a problematic eon to read with any degree of certainty and many of the theories of what may have happened during the first few hundred million years of our planet are based on not much in the way of physical evidence and rather a lot of computer modelling and imaginative thinking. Some bold stratigraphers – risking a severe wag of the finger and a considerable amount of tutting from the International Commission on Stratigraphy – have even attempted to divide up the unofficial eon using geological (or should that be selenological) events on the Moon because there is so little available on Earth.

Much of Earth's early history may be shrouded in mystery but what is certain is that the first 100 million years of the Hadean are well named at least, being a close approximation of hell on Earth. For the rest of the eon, there are a few competing theories that present different views of the exact nature of hell, including one that describes the possibility of a largely frozen Earth with occasional 'impact summers' – warmer spells caused by large asteroid strikes. Another view – one formed relatively recently – is of an Earth of blue skies and wide oceans; a picture more evocative of California without the sports utility vehicles than the traditional fire and brimstone of Hades.

The more familiar scene is one, perhaps, that is foremost in the popular imagination – a bilious volcano of a planet vomiting hot rock everywhere until it cools enough to form a solid surface. This was the initial state, at least, of the Hadean Earth after the impact of Theia – both planet and planetoid are believed to have been molten at the time of the strike and certainly would have been from the moment of impact on. A globe of molten rock, or magma, where, in a well-understood process known as differentiation, heavier elements like iron and nickel sank through it to form the core, while lighter elements like silicon, magnesium and aluminium rose to the top, eventually forming thin fragments of crust. At first, these fragments were similar to the skin that forms around running lava and would have been easily reabsorbed by the magma, only growing more durable as the planet cooled further.

Gases that were present in the magma – like those found in lava – bubbled up through the molten rock and were either lost to space or added to the atmosphere, an atmosphere we would not recognise as 'air' as it was almost completely devoid of oxygen. Earth has only had an oxygen-rich atmosphere for around half its life and during the Hadean eon our planet existed in a poisonous bubble composed predominately of nitrogen, hydrogen, ammonia, carbon dioxide, water vapour and possibly methane, a mix similar to that found on Titan, the second-largest moon of Saturn, today. After perhaps 150 million years of what scientists call 'outgassing' from the molten innards of the planet, atmospheric pressure had also grown very high – estimates range from 20 to 480 times that of modern Earth.

At this point, a hypothetical human (a real one would burst into flames even before they had time to suffocate, be crushed like a beer can or struck by a small asteroid) standing on a thin raft of hot, solid crust would have had quite a view. It would be something like a Roger Dean poster or an album cover for a mid-Seventies prog-rock band, featuring an ocean of bubbling, belching boiling rock upon which floated small pieces of cooled crust, bolides raining from the sky and a rather large and striking Moon, only 14,000 miles away and still on fire.

The view of that Moon may well have been obscured, however, by a primordial cover of dust, smoke and water vapour belched out by volcanic eruptions. The cover would at first have been extremely thick and the atmosphere too hot to condense the vapour to rain. Eventually, as the planet cooled, so the water vapour would cool, first into clouds which then, after further cooling would condense into rain, finally forming early oceans on the crust. We are probably not capable of the imaginative leap necessary to visualise the storms of the Hadean eon, but some scientists believe that annual rainfall was in the magnitude of 157–275 inches (4–7 metres), a continuous torrential downpour for several million years over the entire planet. Today, Earth's fiercest weather comes from temperature differences of a few dozen degrees. Imagine the ferocity of storms that derive their energy from a radioactive planet shrouded in superheated steam.

Scientists used to believe that the planet took the whole of the eon to cool and that a fiery hell was more or less a permanent fixture for the entire length of the Hadean. Some recent discoveries, however, have found some rocks of a Hadean age – just over 4 billion years old – in Canada's Northwest Territories, as well as 4.4-billion-year-old zircon crystals that provide compelling evidence for the presence of surface water at the time of their creation.

For the purposes of unfolding the Hadean past, zircon is a very useful mineral, in that not only are scientists able to measure its age with some precision, but it is also incredibly stable even when subjected to very high temperatures. These findings, amongst others, have led scientists to reappraise the Earth's early days and the vision of a fiery, prog-rock album cover of a planet – a planet that takes a billion years to cool down – is now being replaced by the Cool Early Earth theory. If the Cool Early Earth model is to work, however, a comparative absence of Hadean rock has to be accounted for. A cool Earth would surely leave behind some traces of the geological processes we see today, so why is there a sudden relative profusion of 3.8-billion-year-old rock at the boundary between the Hadean and the following Archean eon, yet only a few occurrences (at least, discovered so far) of older rocks and minerals?

If there are scarce remains of Hadean geology on Earth in comparison to later rocks, the Moon displays the exact reverse – most of its geological evolution occurred during the first billion years of its existence, hence its use as the unofficial marker of Hadean time. Our Moon, now 250,000 miles or so away, has been geologically inactive for a long time and so most of it is made from rock that was last crystallised in the Hadean and Archean eons. Because there is no atmosphere, even the footprints of the astronauts who landed on the Moon forty years ago will still be there and the only weathering that has occurred for most of the rest of its history has been from the impact of small meteorites that have produced the lunar 'soil' or regolith – a process known as 'micrometeorite gardening'. Given that it has all its original geology in place, the Moon provides a number of clues about the Hadean Earth, including evidence that indirectly supports a cooled down Earth with abundant surface water. Not only that, but the Moon may also hold clues to what happened to all of Earth's Hadean geology.

From 4 billion years ago, say the supporters of the Cool Early Earth theory, to the end of the Hadean, the Earth suffered a massive cataclysm known as the Late Heavy Bombardment. We only know of the bombardment because the Moon's large craters appear to have been formed predominately over this period. Samples of Moon rock that were created as a result of large collisions with asteroids and comets were brought back on the *Apollo* missions while others were delivered free to Earth as lunar meteorites. When they were dated, the age obtained suggests that craters on the Moon and, by implication, the Earth, were formed at the hands of an intense attack towards the end of the Hadean eon. The result of a cataclysm the size of the Late Heavy Bombardment – involving tens of thousands of large bolides creating impact craters on Earth anywhere from 12 to 3,000 miles across – would be to remelt the planet's surface afresh until the bombardment stopped, drawing a veil over geological history under which almost all evidence, save for a handful of heat-tolerant minerals, was obliterated.

The Late Heavy Bombardment is only a theory, albeit one with evidence that is plain to see on our closest neighbours in the Solar System, but unless we commit ourselves to an expensive programme of further rock sample collection from the Moon, Mars, Mercury or Venus, we may never know which particular vision of hell applied to the early Earth or for how long.

Another recent theory, however, paints a different picture of the hell on Earth, one that attributes the destruction of Hadean geology to primeval acid rain. Though now largely ignored in the popular consciousness in favour of the more dramatic apocalypse scenario offered by global warming, acid rain briefly became infamous during the 1970s and 1980s for its ability to cause widespread damage to forests and lakes downwind of industrial areas, as well as buildings made of limestone. Rain has always been a weak acid, even before pollution began to be an issue. The mixing of atmospheric water with carbon dioxide creates carbonic acid, a liquid which can over long periods of time dissolve rock (especially alkaline rocks such as limestones and chalk). Man-made acid rain is a more destructive phenomenon – it is formed when waste gases from the burning of fossil fuels, principally sulphur and nitrogen oxides, mix with water vapour in the atmosphere to form dilute acids which then fall as rain. Despite its low profile, acid rain is still having a profound effect on the environment – some rains in industrial areas have been measured to have higher levels of acidity than vinegar.

Back in the Hadean, however, acid rain might have been much, much more corrosive. Estimates of levels of atmospheric carbon dioxide in the Hadean are up to 10,000 times today's value. Aside from the greenhouse effect of a more energetic atmosphere, the acid rain may have been strong enough to dissolve rocks – even resistant ones like granite – rapidly, destroying all evidence of any early continental crust.

The story of the Hadean that emerges from a sampling of these theories is one of a cycle of temperate continental formation punctuated by either occasional massive impacts, some big enough to boil away the oceans, or a rapid and concerted dissolution of the landscape by acid rain. The stretches of temperate conditions in the Hadean may have been for long enough periods for either the development of life or the development of organic chemistry necessary for the later development of life, over 500 million years sooner than was previously thought. Any organisms that did begin to develop during the Hadean would be micro-organisms – life of the kind that doesn't fossilise well even when there are rocks to investigate.

We may never know for sure, but a chemical signature

in 3.8-billion-year-old rocks in Greenland currently holds the record for the earliest sign of life, although it is still a matter of debate. One problem with the signatures is their punctual timing – it suggests that as soon as the hostile conditions of the Late Heavy Bombardment were over, life just spontaneously appeared. This somehow seems less likely than an alternative view: that life was already present in isolated, sheltered niches but only blossomed when the bombardment ceased, like someone waiting in a shop doorway for the rain to stop. With the window of opportunity widened by 600 million years – back to the Cool Early Earth indicated by those 4.4-billion-year-old zircon crystals – conditions may have already existed for long enough for the organic chemistry to assemble itself into self-replicating organisms of some kind.

But what of Britain? Is it possible that some part of one of those initial rafts of cooling rock is forever Hemel Hempstead, Lochinver, Leicester or Llandudno? Probably not, except in the sense that all of that initial crust – the lightweight scum that floated on the magma oceans – was recycled sooner or later and its constituent molecules could now be anywhere. Most of the matter that existed on Hadean Earth is still here, from the inner core of the planet, a solid iron-nickel alloy ball bearing about 1,500 miles wide, through the outer core, the mantle and out to the crust. All of us, unless you consider that we have extraterrestrial origins, are also made of the same stuff, endlessly reincarnated since the Hadean; there are bits of you that might be early crust and so hell, in one form or another, is now probably thinly spread over the whole of Earth.

Further Reading

I never let my schooling stand in the way of my education.

Mark Twain

Geology and landscape can be necessarily complex subjects, but there are plenty of books and websites to help non-scientists unravel it. Your first stop should be this book's companion website, the official home of the British Landscape Club at www.britishlandscape.org, where you will find a chapter-by-chapter breakdown of the book with photographs and links to further information. All of the links to websites listed in this appendix or resources highlighted in the text are also there to save you the bother of manually entering the addresses found on these pages. You will also find Amazon, Abe Books and Waterstone's links to all the books listed below.

Beyond the bubble that surrounds this volume, there are many excellent resources out there, of which I have selected a few here. Starting with the books, I can heartily recommend a couple of volumes which should cover all the information you will need for a light skimming of the subject in hand.

Books

The Hidden Landscape by Richard Fortey. At the time of writing, this award-winning book is out of print but a new edition is apparently forthcoming. In the meantime, you might be able to pick up a second-hand copy of this classic dissection of the British countryside. Marvellous reading material told with equal measure of enthusiasm and insight.

The Geology of Britain: An Introduction by Peter Toghill. A full-colour and friendly A4 textbook for the lay person as well as students: gloriously illustrated with excellent diagrams and maps, but highly technical and detailed in places. It might be your next step.

Geology and Scenery in England and Wales by A.E. Trueman. Originally published as *The Scenery of England and Wales* in 1938, this Pelican imprint first came out in 1949 with reprints up to 1963, so some of the details have since been superseded by more modern geological analysis. It eschews the timeline treatment of the landscape in favour of an area-by-area analysis, so it is a useful volume to have around if you're thinking of exploring another area on your holidays.

The History of the Countryside by Oliver Rackham. Also available in a glossy and lavishly illustrated edition, this is the seminal treatise on the human history of the British countryside. Perfect if you want to explore the history of woodland, field boundaries, roads and paths.

Be Your Own Landscape Detective by Richard Muir. In his illustrated guide to human landforms Muir takes a more workbook-like approach, guiding the reader through different kinds of landscapes.

Local Guides

Your local council may publish a free guide to local geology for tourists. These are often rather good and easily worth the journey to a local tourist information centre. There may also be other county guides available, written by local authors or groups that can guide you to interesting spots in your neighbourhood.

Scottish Natural Heritage publish a series of glossy booklets, each highlighting a part of the Scottish geological heritage. Some of these may even be downloaded for free from their website.

Maps

The British Geological Survey (BGS) produce a range of maps at different scales for a multitude of different uses and users. By far the most useful – if you want to explore your local landscape in relative detail – are the 1:50 000 series (the same scale as the pink covered 'Landranger' maps of the Ordnance Survey) which cover virtually the whole country in 20 × 30 kilometre tiles. For some peculiar reason, they are landscape in orientation in England, Wales and Northern Ireland, but portrait format in Scotland. The BGS also sell small-scale poster maps of the UK and the British Isles which simplify the geology according to the demands of the scale.

Internet

As well as the usual places you gather information from on the Internet there are many specialist sites which can be helpful in your landscape detective work. Websites come and go, however, and you will find an up-to-date list on www.britishlandscape.org. There are also a few surprises on the Internet. You can, for instance, get free introductory-level Open University videos from iTunes University (not only on geology and geography, but on many other subjects), for which you will need the free iTunes program – just click on iTunesU in the iTunes Store. I will endeavour to keep an up-to-date link on www.britishlandscape.org because of the ever-changing nature of the iTunes Store.

The British Geological Survey, the government body with reponsibility for mapping the country's geology, has many resources for downloading, including a free Google Earth overlay that roughly corresponds to the level of detail of the poster maps.

Even without the geological overlay, I can thoroughly recommend the free version of Google Earth if you don't already have it, simply because it brings detailed satellite photography to the computer desktop. With Google Earth, you will be able to participate in *The Lie of the Land* fly-by and visit some of the locations in this book virtually. Full details of where to obtain Google Earth, the BGS overlay and *The Lie of the Land* fly-by are available on the website www.britishlandscape.org.

Glossary of Terms

alluvium – Any deposit laid down by a river.

anticline – An upfold of sedimentary rocks where the oldest rocks are found at the core.

arthropod – Invertebrate animals with an exoskeleton. Includes crustaceans, spiders, insects and trilobites, amongst many others.

arête – A knife-like ridge of rock which separates two U-shaped glacial valleys on a mountain.

basin – A bowl-like syncline; a depression of strata formed by tectonic forces, where the youngest rocks are at the centre.

batholith – An accumulation of plutons; cooled intrusions of magma.

calcite – A mineral formed from calcium carbonate and making up the majority of rocks such as limestone or chalk.

cirque, corrie, cwm – An ampitheatre-like bowl formed at the head of a glacier.

crust – The outer solid shell of the earth.

cuesta – An alternative name for a gentle escarpment.

deposition – The addition of sedimentary material on the surface of the landscape.

dip slope – The shallow slope of an escarpment, so-called because it follows the angle of dip of the underlying rocks.

dome – A roughly circular form of an anticline, an upturned bowl of rock strata where the oldest rocks are at the core.

escarpment – Often interchangeable in meaning with a scarp slope, colloquially and in this book referring to a long ridge consisting of a scarp and a dip slope.

fault – Where two large blocks of rock move relative to one another, or the line where this happened.

igneous – Rock formed from cooled magma or lava.

lava – Molten rock that has reached the surface at the eruption of a volcano.

magma – Molten rock that has not reached the surface.

mantle – The interior of the earth between the crust and the core.

metamorphic – Rock recrystallised in situ by intense heat and/or pressure.

metamorphism, contact – The process of metamorphosis by intense local heating, such as the proximity of a magma intrusion.

metamorphism, regional – The process of metamorphism over a wide area, by tectonic activity or simply as a result of being at great depth in the crust.

mya – A convention used in geological texts to abbreviate 'millions of years ago'.

orogeny – The process of mountain building through the compression of part of the earth's crust.

pluton – An intrusion of magma that forms a body of igneous rock.

protolith – The original rock from which a metamorphic rock was derived.

quartzite – A metamorphic rock which was originally sandstone.

rocks and minerals – A rock is an aggregate of minerals. Minerals are naturally occurring products of geological processes which have a specific chemical composition.

scarp slope – The steep slope of an escarpment or cuesta.

sedimentary – Rock formed from sediment that has been deposited.

silica – Silicon dioxide, the most abundant mineral in the earth's crust and a major constituent of sandstone.

strata – Layers of sedimentary rock.

stratigraphy – A branch of geology concerned with the order and relative position of strata and their relationship to the geological timescale.

syncline – A depression of strata formed by tectonic forces, where the youngest rocks are at the core.

terrane – A block or fragment of the earth's crust with its own distinct geological history. It's 'exotic' history is a result of an origin from a different crustal plate from the one where it is found.

vacuole – An empty space within a rock, formerly occupied by a gas.

volcanic island arc – A curved chain of volcanic islands above the margin of two converging plates, where one plate is being subducted under another. There is typically a deep ocean trench on the convex side.

Geological time

eon – The longest formal unit of time on the geological timescale; there have only been four eons since the earth was formed: Hadean (unofficial, from 4.6 billion to 4 billion years ago); Archean (4 billion to 2.5 billion years ago); Proterozoic (2.5 billion to 542 million years ago); and Phanerozoic (542 million years ago to the present day). All but the Phanerozoic are in the Pre-Cambrian super-eon.

era – Subdivision of an eon; there have been ten eras since the end of the Hadean eon, which is not subdivided into eras.

system/period – Subdivision of an era; there have been twenty-two periods since the end of the Archean eon, which is not subdivided into periods. Except for a few cases, the time span of one of the chapters in this book is a period. 'System' refers to the geology of a period.

series/epoch – Subdivision of a period; there have been thirty-eight epochs since the start of the Cambrian period/Palaeozoic era/Phanerozoic eon, 542 million years ago. Anything older is not subdivided into epochs. 'Series' refers to the geology of an epoch.

age/stage – Subdivision of an epoch. 'Stage' refers to the geology of an age.

Acknowledgements

I'd like to thank all the fantastic people who have helped me write this book in some way. Naturally that not only includes my editor Bruno Vincent, whose input improved my output immeasurably, but also Jon Butler who commissioned the book in the first place and my agent and friend Simon Benham.

Thanks to the Vince girls – my wife Kate, our daughter Freya, plus one other who will be with us by the time you read this – who continue to put up with this kind of thing all the time. Thanks also to friends – Tom Bromley, Neil Smith, Jon Bayliss and Matthew Thomas, who furnished just the right amount of beer and advice and to Pat Shelley for the guided tour and for furnishing beer in exactly the wrong amount.

There are many others. From Donald the postman who was so obliging with his time and stories as we drove through the snowy Highlands, to the potty-mouthed and apparently avi-phobic birdwatcher I met on the River Teme outside Ludlow, shooing off a mallard and swearing at everything in the sky, including a hang glider. They all go to prove that if you scratch the surface of this country, you'll always find a gem.